Careers in Chemical and Biomolecular Engineering

Careers in Chemical and Biomolecular Engineering

Victor H. Edwards
Suzanne Shelley

CRC Press
Taylor & Francis Group
Boca Raton London New York

CRC Press is an imprint of the
Taylor & Francis Group, an **informa** business

CRC Press
Taylor & Francis Group
6000 Broken Sound Parkway NW, Suite 300
Boca Raton, FL 33487-2742

© 2019 by Taylor & Francis Group, LLC
CRC Press is an imprint of Taylor & Francis Group, an Informa business

No claim to original U.S. Government works

Printed in Canada on acid-free paper

International Standard Book Number-13: 978-1-138-09991-3 (Paperback)
International Standard Book Number-13: 978-0-8153-8086-3 (Hardback)

Library of Congress Cataloging-in-Publication Data

Names: Edwards, Victor H., author. | Shelley, Suzanne, author.
Title: Careers in chemical and biomolecular engineering / Victor Edwards, Suzanne Shelley.
Description: Boca Raton : Taylor & Francis, CRC Press, 2018. | Includes bibliographical references.
Identifiers: LCCN 2018007562| ISBN 9781138099913 (pbk. : alk. paper) | ISBN 9780815380863 (hardback : alk. paper)
Subjects: LCSH: Chemical engineering--Vocational guidance. | Biochemical engineering--Vocational guidance
Classification: LCC TP186 .E39 2018 | DDC 660.023--dc23
LC record available at https://lccn.loc.gov/2018007562

Visit the Taylor & Francis Web site at
http://www.taylorandfrancis.com

and the CRC Press Web site at
http://www.crcpress.com

To my beloved wife, Mary, and to our two dear children, Hal and Beth, and their families; to my brothers, Dave and Tim; to my chemical engineering colleagues and to many other great friends for their insights, wisdom, companionship, and help when needed. And in memory of my late parents (Philip and Margaret Edwards).

The wind flows free

That's as it should be

Stay as free as the wind

And go forward my friend

Victor H. Edwards

To my husband and daughters, Bruce, Samantha, and Charlotte Margolin, and my siblings, Stef Shelley, Matthew Shelley, Meredith Shelley McCracken, and Jan Phillips; to those close friends who provided creative, technical, and moral support during the writing of this book; and in memory of my beloved, always-supportive, and inspirational parents, Gabrielle and Joseph Shelley.

Such wealth

Your sweet love brings

I scorn to change my state

With kings

Suzanne Shelley

Vic and Suzanne would also like to offer a most heartfelt thank you to each of the 25 pre-eminent chemical and biomolecular engineers profiled here—Some of whom were already friends or professional colleagues of ours, and some of whom were recommended to us by others and have since become our friends. Each one of them spent many hours on the phone with us. Each one of them shared their memories and insightful stories of personal and professional satisfaction and challenges, and offered their wisdom and "best practices" for engineering students and young engineers, with humor and grace. This book has been greatly enhanced thanks to the collective input of these talented individuals, whose voices are representative of the industries and the career options we have worked to describe for young engineering students.

Contents

Profiles of Working Engineers

Preface for Students, Guidance Counselors, Mentors, and Teachers

CHEMICAL AND BIOMOLECULAR ENGINEERING: A PROUD PAST AND A BRIGHT FUTURE

The fields of chemical and biomolecular engineering are both broad and dynamic, offering students a seemingly endless array of career opportunities and pathways for intellectual and practical satisfaction. These careers are often aimed at satisfying society's need for safe; sustainable energy; plastics, metals, alloys, and other advanced materials; innovative pharmaceuticals; high-performance building materials; specialty chemicals; paper, agricultural and forest products; and a cleaner environment.

In the past, most chemical engineering students had their first exposure to chemical engineering courses by their second or third year of college. This is because of the important need for engineering students to first establish a foundation in basic science, mathematics, and the humanities. During the college years, students who were pursuing a chemical engineering degree were given a step-by-step, thorough grounding in increasingly sophisticated chemical engineering principles. The curriculum historically emphasized the theoretical aspects of chemical engineering science—rather than the practical aspects of chemical engineering. In the past 50 years, college courses typically provided very little introduction to the many different industry sectors and career paths that would be available to those earning a chemical engineering degree.

The Capstone Plant Design Course in the fourth or fifth year involves groups of four or five students conducting a year-long design of a hypothetical process plant. This was traditionally the first time that students could clearly see how the preceding math, chemistry, engineering, and related courses were woven together. This sequential and hierarchical approach to the chemical engineering curriculum ultimately provides students with a solid foundation in the fundamental principles of chemical engineering—but often leaves them wanting when it comes to truly understanding how a chemical or biomolecular engineering degree may be put to good use in the many different industry segments that depend on such graduates.

To gain potential career insights, most students typically depend on summer or co-op jobs, the news media, and anecdotes from friends and faculty who may be familiar with these two dynamic disciplines or the industries that value them. At best, those opportunities may be relatively slender in scope, offering a narrow glimpse into one opportunity or another, rather than providing a thorough understanding of the many professional avenues, across many different industrial sectors, and the many fascinating career options that are available to chemical engineers and biomolecular engineers. The many opportunities that are available for such graduates allow them to both make demonstrable contributions to society and enjoy lucrative, satisfying careers.

GROWING OPPORTUNITIES

The scope of opportunities in chemical and biomolecular engineering has grown tremendously over the past few decades, and today chemical engineers are able to find interesting, enriching career opportunities in an ever-widening array of industries and applications. Students tell us that now, more than ever, that there is a great need for a concise book that explores and explains the chemical and biomolecular engineering practice in depth, and describes the many *career paths* that are available for students who are willing to undertake these relevant, challenging majors. This book is written to inform students who are pursuing a chemical engineering degree of the many opportunities and to help them to more meaningfully explore some potential early career decisions.

Several excellent books have already been published to support freshman courses in chemical engineering, but these books tend to rely on simpler chemical engineering calculations to illustrate the practical application of this course of study. Many universities still begin teaching in-depth chemical engineering courses—which gives students a better idea of the breadth and depth of the industries and career options that are available to chemical and biomolecular graduates—only later in their curriculum. And many of these books emphasize theory, calculation methods, unit operations, and chemical engineering principles—all essential aspects of the curriculum—but most provide little if any discussion on the scope of *employment and career opportunities* that may be available throughout the global chemical process industries (CPI). This is especially true about the many other newer areas of opportunity for working chemical engineers, such as the fields of biotechnology and nanotechnology. *This book aims to fill that need.*

ABOUT THIS BOOK

This book outlines the past contributions of chemical and biomolecular engineering to society, highlights the breadth and depth of current opportunities, and forecasts future contributions of working chemical engineers to the modern society we have come to enjoy. Chemical and biomolecular engineering concepts are described in a way that can be understood by high school and early college students, and the adults (parents, teachers, school advisors, guidance counselors, and other mentors) who may guide them.

PROFILES IN CHEMICAL AND BIOMOLECULAR ENGINEERING

In addition to efforts to describe the many diverse industries and career pathways that are available for chemical engineering and biomolecular engineering graduates, this book also aims to "humanize" these dynamic professions. Toward that end, this book also provides interview-style articles that profile 25 industry-leading chemical and biomolecular engineers. These Profile articles share first-person narrative reflections and insights based on their own career trajectories. Specifically, each of these 25 Profile articles provides the following insight, shared by each seasoned chemical engineer:

- A biographical sketch
- A brief outline and shared anecdotes to illustrate how their chemical engineering education has translated into enjoyable, rewarding career experiences
- A "look ahead," as each of these professional chemical engineers shares insight into what the future might hold within their particular fields and within the realm of chemical engineering in general
- Words of advice, recommendations, and best practices for students of chemical engineering to pass along some of the wisdom gained from their career and life experience

THIS BOOK PROVIDES GUIDANCE

This book offers high school and college students, parents, teachers, and guidance counselors with an easily accessible overview of the many roles that chemical and biomolecular engineers play in our complex society. Such roles include research, development, design, and operation of industrial processes throughout the diverse *chemical process industries.*[*] Similarly, it has been shown time and time again that a degree in chemical and biomolecular engineering also provides a solid and valued background for other professional career pathways, such as intellectual property (patent) law, energy economics, biomedical research and engineering, pharmaceutical development, process safety, and energy management.

In addition to serving as a reference book, this book will also serve well as a supplement to both introductory courses on chemical engineering theory and calculations, and other "introduction to engineering" college courses that are aimed at helping students to decide which branch of engineering (and thus which course of study) might be most interesting to them.

Victor H. Edwards
Suzanne Shelley

---

[*] The *chemical process industries* (CPI) represent, by definition, many different but related industry sectors; these include, but are not limited to, chemicals production, petroleum engineering, pharmaceutical manufacturing, energy and semiconductor manufacturing, food and beverage manufacturing, environmental management, and many more.

Authors

Victor H. (Vic) Edwards, PhD, PE is a chemical engineering consultant based in Houston, Texas. Before his retirement, Vic practiced for 30 years at IHI Engineering and Construction International and predecessor engineering companies. In his most recent role at IHI, Vic served as director of process safety. He was responsible for health, safety, and environment (HSE) on many major projects—three as large as US$3 billion. Clients included ExxonMobil, Dow Chemical, Dominion Resources, DKRW, Alcoa, Kennecott Copper, Lyondell, and Chevron. Vic and his team received the Process and Construction Award for HSE Excellence in Design in 2008. Prior to that, Vic was process director; in that role, he led the process engineering, environmental engineering, and process safety management as a DuPont engineering alliance contractor. During that time, Vic contributed to numerous small DuPont capital projects at ten Gulf Coast (USA) plants. He and members of his team received three awards from DuPont for safety and environmental excellence in engineering design, and two more DuPont awards for safety and environmental excellence. In 1998, Edwards was named Employee of the Year.

During the first 20 years of his career, Vic held faculty positions at Cornell University (Assistant Professor of Chemical Engineering, 1967–1973) and Rice University (Visiting Professor of Environmental Engineering, 1979–1980), and he spent a year at the National Science Foundation (on leave from Cornell in 1971, to help start the first national program on biotechnology). Early in his career, Vic devoted three years (1973–1976) to pharmaceutical research as a research fellow at Merck. He also led the creation of a US$12.8 million project related to research, development, and demonstration of a novel process to produce synthetic natural gas from aquatic and terrestrial biomass and municipal carbonaceous wastes for United Gas Pipeline. (This work earned four patents, from 1976 until 1979.)

Vic is the 2015 recipient of the Norton H. Walton/Russell L. Miller Award from the Safety and Health Division of the American Institute of Chemical Engineers (AIChE). The award recognizes outstanding achievements in, and contributions to, chemical engineering in the areas of loss prevention, safety, and health.

In 2013, he chaired the AIChE 9th Global Congress on Process Safety, the first Process Plant Safety Symposium in 1992, and the Second International Plant Operations and Design Conference in 1997. He has also served on technical advisory committees for the American Petroleum Institute, DuPont Engineering, Prairie View A&M University, Rice University, and Texas A&M University. He has received service awards from Prairie View and Texas A&M Universities. Vic has published more than 60 technical papers and several book chapters, and he edits Section 10 of *Perry's Chemical Engineers' Handbook*.

Vic holds a B.A. degree from Rice University and a PhD from the University of California at Berkeley, both in chemical engineering. He is a registered professional engineer in Texas, and a Fellow of the American Institute of Chemical Engineers (AIChE). He is also a member of the American Association for the Advancement of Science, the American Chemical Society, the National Fire Protection Association, the National Society of Professional Engineers, and the New York Academy of Sciences.

Suzanne Shelley has been an Editor at *Chemical Engineering* magazine (originally published by McGraw-Hill, now owned by Access Intelligence) since 1989. She served as a full-time editor at the magazine from 1989 until 2005 (and was the magazine's Managing Editor from 2000 until 2005). She left the magazine in 2005 to launch her own company, Precision Prose Inc., to provide freelance technical writing, editing, and ghostwriting services to a diverse array of companies and magazines in the chemical process industries (CPI) and pharmaceutical sector. In that capacity, Suzanne has remained a regular freelance contributing editor at *Chemical Engineering*, and she also serves as a regular freelance contributing editor at *Pharmaceutical Commerce* magazine. For several years, Suzanne also served as a freelance contributing editor at *Chemical Engineering Progress* magazine (CEP; AIChE) and *Turbomachinery International* magazine. Through her consultancy, Suzanne provides engineering- and business-oriented writing, editing, and ghostwriting services to numerous operating companies, third-party service providers and contract research organizations (CROs), consulting companies, and marketing/advertising agencies. During the 1990s, she launched *Environmental Engineering World* magazine as a sister publication to *Chemical Engineering* and served as its Executive Editor from 1992 until 1996. In 2005, Suzanne authored the Introduction chapter (70 pages), entitled "Nanotechnology: Turning Basic Science into Reality," to the book *Nanotechnology: Environmental Impacts and Solutions* (Lou Theodore and Robert Kunz, John Wiley, 2005). Suzanne was also one of the "Happy 100" women profiled in the book *Be Happy at Work: 100 Women Who Love their Work and Why* (Joanne Gordon, Ballantine/Random House; the paperback version of this book was later re-titled *Career Bliss: Secrets from 100 Women Who Love Their Work*). She holds a BS in Geology (Honors) from Colgate University and an MS in Geology from the University of South Carolina (Columbia). Prior to joining *Chemical Engineering* magazine, Suzanne worked as a deepwater exploration geologist for Amoco Production Co. (New Orleans) in the 1980s. Between college and graduate school, she worked for the US Geological Survey in Golden, CO, on the US government's Nevada Test Site Seismic Risk Assessment Project.

What Are Chemical Engineering and Biomolecular Engineering?

"Scientists investigate that which already is; engineers create that which never has been."

Albert Einstein

Loosely paraphrased: "Science is what is... Engineering is what you want it to be."

SCIENCE AND ENGINEERING: TWO ESSENTIAL AND RELATED DISCIPLINES

Chemists study matter and its transformation into its different forms, and create new chemicals, including chemicals of great potential value. They also devise ways to synthesize those chemicals; first, they do this in laboratory-scale quantities, often with the help of new catalysts. Then chemists work with others (including chemical engineers and those in related disciplines) to scale up these discoveries to produce commercial-scale quantities using processes that are deemed safe and environmentally and economically viable. Chemists are increasingly seeking to apply principles of "green chemistry" the practice of creating new chemicals and chemical reactions that are benign in terms of their ultimate impact on personnel and the environment.

By contrast, chemical engineers play an integral role in taking promising chemistry-related discoveries and devising the necessary engineering, safety, and control systems to produce them on a commercial scale, both safely and economically, with minimal impact to the environment. Such professionals use the full range of chemical engineering principles to design, build, and manage industrial-scale process plants that are needed to safely and sustainably produce the needed quantities of useful chemicals and related products for society with minimum impact on the environment. Figure 1.1 shows several chemical plants on the coast of the Netherlands.

Chemical engineers often work closely with chemists to take promising breakthroughs and produce commercial-scale quantities of the new products and processes.

Similarly, biochemists, microbiologists, and other life scientists study how living organisms synthesize and metabolize biochemicals; they also study ways to create useful biochemicals at laboratory scale and then work with biomolecular engineers to devise ways to produce useful bioproducts safely, economically, sustainably, and at commercial scale. Figure 1.2 shows a pharmaceutical manufacturing facility.

BENEFITS OF CHEMICAL AND BIOMOLECULAR ENGINEERING

Have you ridden in a car recently? If so, you can thank chemical engineers for the gasoline, rubber for the tires, plastic for the steering wheel and interior, safety glass for the windows, for the microchips, the catalytic converter to minimize air pollution, and many other components.

Chemical engineers are now creating fuel cells to power cars using cleaner-burning hydrogen produced by other chemical engineers. Would you like to become one of those chemical engineers? Or would you like to become a chemical engineer working on new batteries that would extend the range of electric cars and reduce the risk of fire?

FIGURE 1.1   This aerial view shows several chemical process plants in the Netherlands. Chemical engineers and biomolecular engineers find a diverse array of employment opportunities at these types of facilities, which rely on all aspects of these technical disciplines to transform raw materials into finished products in a safe, efficient and economical way. (Source: Shutterstock ID 717938881, 7 August 2017, Moerdijk, Holland. Aerial view of industrial area with Shell Chemicals Attero and Essent. On the clear horizon a sky with clouds and river Hollands Diep)

FIGURE 1.2   Pharmaceutical manufacturing facilities place a high premium on ensuring high standards of purity when manufacturing active pharmaceutical ingredients and final pharmaceutical products. (Source: Shutterstock ID 214256035, pharmaceutical factory equipment mixing tank on production line in pharmacy industry manufacture factory)

When you take medicine, you can thank chemical and biomolecular engineers for devising ways to manufacture the various medications (in pill form, as an injection, or via an intravenous drip) in ultra-pure, ultra-clean (aseptic) facilities, at a production scale that is sufficient to be both practical and economical. These highly trained technical experts continue to ensure high purity in pharmaceutical products by devising highly specialized packaging and administration routes (intravenous and injection systems, blister packs for pills and capsules, and more). Biomolecular engineers are also helping to discover new antibiotics, new cancer medications, treatments for rare diseases, and so much more, and then to produce them, purify them, and package them—all on an industrial scale.

TABLE 1.1   Some Great Achievements of Chemical Engineers

Semiconductor fabrication

Medicine

Environmental protection

Crude oil (petroleum) processing

Plastics

Synthetic fibers

Synthetic rubber

Gases from air

Food

Antibiotics

Separation and use of Isotopes

Source: AIChE, *Chemical Engineering Progress*, November, 5–25, 2008.

Chemical and biomolecular engineers routinely make enormous contributions to society, in many different industrial segments (Table 1.1).

CHEMICAL AND BIOMOLECULAR ENGINEERING DEFINED

Chemical engineers work with chemicals such as ethylene and polyethylene; biomolecular engineers work with (bio)chemicals of biological origin, such as insulin and penicillin.

The professions of chemical engineering and biomolecular engineering are very similar; they use many of the same processing methods to create industrial- or commercial-scale quantities of new chemicals, biochemicals, medicines, fuels, specialty materials, and more. In fact, some believe that the name "chemical engineering" covers all types of materials. We will maintain the distinction between chemical engineering and biomolecular engineering because there are a few important differences in the processes employed across these two disciplines. Figure 1.3 is a photo of two engineers reviewing plans in a process plant.

Chemical engineering. This discipline applies chemistry, physics, mathematics, and unique chemical engineering principles to create and manage industrial processes that transform raw materials into useful

FIGURE 1.3   Chemical engineers often pursue satisfying technical and managerial careers working in facilities that produce petrochemicals, plastics, and related products. When such facilities use biological processes to carry out the desired chemical conversions or to treat waste streams, biomolecular engineers are also in demand. (Source: Shutterstock ID 143754721, chemical engineering manager and senior worker at petrochemical plant)

products safely in a sustainable way. These transformations typically involve changes in the physical state, chemical composition, and other characteristics.

Biomolecular engineering. In addition to using many of the chemical engineering processes, biomolecular engineering also employs biological and biochemical sciences to develop processes to produce biomolecular products (such as medicines, fuels, amino acids, and vitamins) using unique chemical and biomolecular engineering principles. These bioproducts often involve the direct use of enzymes, living cells, or plants in the manufacturing process.

A *chemical or biomolecular process.* This process involves a combination of steps in which starting materials are converted into desired products using equipment and conditions that facilitate that conversion (Stolen and Harb 2011).

ROLES OF CHEMICAL AND BIOMOLECULAR ENGINEERS

Chemical engineers play a variety of roles in the chemical process industries; Table 1.2 lists a few examples. We will provide more details later, after briefing on chemical engineering principles. Throughout this book, we also share profiles of 25 outstanding chemical and biomolecular engineers, chosen for the diversity of their careers. They share stories of their career experience and wisdom for young engineering students.

TABLE 1.2   Examples of Professional Career Paths Available to Chemical Engineers and Biomolecular Engineers

University teaching
Basic research
Basic research on chemical and/or biomolecular principles and methods
Process research and development on new or improved processes
Product research and development on new or improved products
Process technology
Process engineering and design
Environmental engineering
Process safety engineering
Health, safety, and environmental management
Plant commissioning and startup
Plant operations and process improvement
Plant management
Corporate management
Government service in organizations such as the Environmental Protection Agency, Occupational Safety and Health Administration, Department of Energy, National Science Foundation, and National Institutes of Health
Engineering consulting
Patent and intellectual property law
Regulatory law
Corporate law
Biomedical engineering

THE GRAND CHALLENGES OF ENGINEERING IN THE TWENTY-FIRST CENTURY

A few years ago, the National Academy of Engineering identified 14 global "Grand Challenges" (Table 1.3). These are areas in which engineers have an important role to play, to improve the quality of life for citizens of the world (US Academy of Engineering 2017). In Table 1.3, the nine specific challenges where chemical and biomolecular engineers can play a particularly important role are designated with the notation "ChBE."

These challenges are global because science and engineering transcend national boundaries, and because these challenges ultimately affect citizens of all countries. The National Academy of Engineering has been organizing periodic international symposia to review progress on these Global Challenges. The Global Grand Challenges Summits (GGCS) aim to spark worldwide interdisciplinary collaborations that will lead to innovative ways of addressing critically important engineering challenges and opportunities and inspire the next generation of change makers.

The third GGCS was held in Washington in 2017 (US Academy of Engineering 2017). The third Global Grand Challenges Summit (GGCS) was jointly organized by the US National Academy of Engineering (NAE), the UK Royal Academy of Engineering (RAE), and the Chinese Academy of Engineering (CAE).

In the next chapter, we summarize several historical advances in chemical and biomolecular engineering and showcase some leading technical professionals who have made them.

INDUSTRY VOICES

"Handling highly flammable, dangerous materials…really keeps you up at night—thinking about all the things that could go wrong, and wanting to always stay ahead of the best-in-class safety tools, training opportunities, automated systems and monitoring devices—to ensure personnel safety, protect the environment and safeguard the company's assets and products."

—John Florez, ExxonMobil, page 8

TABLE 1.3   The Global Grand Challenges of Engineering in the Twenty-First Century[a]

Advance personalized learning (ChBE)

Make solar energy economical (ChBE)

Enhance virtual reality (ChBE)

Reverse-engineer the brain

Engineer better medicines (ChBE)

Advance health informatics

Restore and improve urban infrastructure

Secure cyberspace (ChBE)

Provide access to clean water (ChBE)

Provide energy from fusion

Prevent nuclear terror

Manage the nitrogen cycle (ChBE)

Develop carbon sequestration methods (ChBE)

Engineer the tools of scientific discovery (ChBE)

Source: US National Academy of Engineering, www.engineeringchallenges.org/challenges.aspx and www.nae.edu/MediaRoom/20095/164396/171178.aspx, accessed November 12, 2017, 2017.

[a] Those challenges that show the note "(ChBE)" indicate specific challenges for which chemical and biomolecular engineering can make specific and significant contributions.

HOW DOES ONE BECOME A CHEMICAL OR BIOMOLECULAR ENGINEER?

The minimum requirement to become a chemical engineer is the completion of a four-year bachelor's degree (BS) at a college or university. There are well over 100 ABET-accredited (Accreditation Board for Engineering and Technology) programs in chemical engineering in the United States.* Advanced degrees, such as an MS (Master of Science or Master of Engineering, which typically requires an additional two years of school following an undergraduate degree), or a PhD (Doctor of Philosophy, which typically requires an additional four years of school following an undergraduate degree) provide additional qualifications that offer advantages in roles in process and product technology, research, development, and engineering. Students seeking these advanced degrees often benefit from working in industry for several years after the bachelor's degree before returning to the university for graduate school.

ADDITIONAL QUALIFICATIONS AND CERTIFICATIONS

Just as doctors and lawyers require state licenses, all 50 states and the District of Columbia require licenses to practise engineering in certain jobs and functions. In most US states, the Professional Engineering (PE) license requires the following achievements:

- A bachelor's degree from an accredited college or university
- Passage of the Engineer-In-Training Exam (usually taken right after the BS)
- Four years of documented engineering practice working under the supervision of an engineer who has a PE license
- Passage of the Principles and Practice Exam (typically taken after completion of four years of practice in an engineering role)

Requirements differ from state to state and are readily available on the web. In states that offer the industry exemption, licensure is not required unless the engineer is delivering engineered products directly to the public. Many chemical engineers working for large companies do not pursue the PE license, but the PE license is still a worthy professional goal. Some industries or specific roles will require it, and many engineers who go on to provide consulting work or go into the legal profession value having a PE because it becomes a legitimate, recognized credential to help underscore that person's technical and professional expertise.

Many other certifications exist, some of which are offered by professional organizations. One example is Certified Safety Professional, offered by the Board of Certified Safety Professionals. Another is board certification as a Diplomat of Environmental Engineering, offered by the AAEECB (American Academy of Environmental Engineering Certification Board). Certifications like these are helpful to chemical or biomolecular engineers who are specializing in the corresponding area of professional practice.

---

* The current list can be found on the website of the American Institute of Chemical Engineers (AIChE): www.aiche.org /community/students/career-resources-k-12-students-parents/accredited-programs-chemical-engineering.

ASIDE: ELEMENTARY CONCEPTS IN CHEMISTRY

The forms of matter that we encounter most in everyday life exist in one of three forms: solids, liquids, and gases.

Water is an example of a material that occurs as a solid (ice); as a liquid (water); as a gas (steam), depending on the conditions (such as temperature, pressure, and the presence of impurities) to which it exposed. Examples of atoms or elements include hydrogen, oxygen, carbon, nitrogen, sodium, chlorine, and many more. Many atoms combine with one another to form molecules.

Water is also an example of a molecule; it has the chemical formula H_2O. "H" represents hydrogen; "2" represents the two atoms of hydrogen in a water molecule; "O" represents one atom of oxygen in a water molecule. Each of the two hydrogen atoms is bonded to the oxygen atom. When hydrogen atoms and oxygen atoms are mixed together under the right conditions, they combine, via a process called a chemical reaction, to make water molecules. This reaction releases energy in the form of heat.

The bonds that are produced are strong enough that water is not easily turned back into its constituent atoms—hydrogen and oxygen. Water molecules (and, in fact, all atoms) are very small. When we look at a glass of water, we are seeing liquid that is made up of more than 10^{24} water molecules.

Matter is composed of atoms (or elements)—the basic building blocks—and molecules. Molecules themselves are composed of at least two atoms, while special molecules such as plastics and DNA may be composed of hundreds of thousands to millions of atoms.

Atoms or elements are the simplest form of matter; there are a little over 100 known elements.

- In solids, atoms are in close contact with one another and do not move readily. As a result, most solids are the densest form of matter for that atom or molecule.
- In liquids, the atoms or molecules are in close contact, but they have enough energy to move easily. Most liquids are less dense than the corresponding solid form of the same substance. Water presents an exception to this general rule of thumb, because its solid form (ice) is less dense than its liquid form. This helps to explain why the less-dense ice cubes float, rather than sink, in your glass of ice water.
- In gases, the atoms or molecules are so far apart that they move about easily, and most gases are much lighter than the corresponding liquid form of the same substance. When we see water boiling (in other words, being converted to its gaseous state), we can see the bubbles of steam form in the liquid water and rise to the surface, and then plumes of steam condensate (small droplets of liquid water condensed from steam) rising into the air above the pot of water.

While only about 100 elements have been identified to date, these elements can often combine with one another or with other elements to create an almost-infinite number of chemical compounds. Many molecules can also combine with other molecules and/or elements. The chemical bonds that form between atoms and molecules can produce an innumerable number of possible compounds.

Finding ways to both produce—and then create practical applications for—this infinite array of compounds and the many forms of matter is the precise realm that chemical and biomolecular engineers operate in. Practical applications are often achieved by creating matter in the forms that are most useful, and then employing the full range of chemical engineering and biomolecular engineering principles and techniques to turn a promising discovery into a reality—by designing, operating, maintaining, controlling, and optimizing all of the various engineering, safety monitoring, and control steps that are required in an overall process to safely and economically produce the target compound in desired, commercial-scale quantities. The work of chemical and biomolecular engineers is essential to society and intellectually and financially satisfying to the many men and women who work in the diverse fields that require such expertise. Many of these endeavors are discussed and many satisfied technical professionals are profiled throughout the many chapters of this book, to provide an "up close and personal" view of what "life as a chemical or biomolecular engineer" really looks like.

PETROLEUM REFINING

In his own words
Involved in all aspects of petroleum refining
"from the shop floor to the top floor"

JOHN FLOREZ, ExxonMobil

Current position:
Department Head, Fuels North Operations, ExxonMobil's Baytown Refinery (Houston, TX)

Education:
- BS, Chemical Engineering, New Mexico State University

A few career highlights:
- President of Houston Mexican-American Engineers and Scientists (MAES) Professional Chapter
- Position on MAES National Board of Directors, 2007–2012
- President of the ExxonMobil Baytown Refinery Global Organization for the Advancement of Latinos (GOAL) employee group
- Profiled in *Chemical & Engineering News* magazine, October 2015

John Florez has spent nearly two decades working at ExxonMobil—his entire chemical engineering career to date—moving through a series of increasingly challenging roles, each demanding and honing a different set of skills and responsibilities. And he wouldn't trade his experience for anything.

In keeping with ExxonMobil's corporate philosophy of rotating its engineering leaders through a series of assignments, Florez has held a diverse array of positions, both in the US and abroad, moving to his next assignment roughly every two years. "Being thrown into new, challenging situations is always hard, and it always takes a while for me to say I actually feel proud and competent and confident in my new role," he says. "However, the company is doing this on purpose—to constantly challenge me to develop deeper and broader skill sets, and to expand my world view and familiarity with its complex global operations."

Florez is currently the Department Head for the Fuels North Operations at ExxonMobil's Baytown Refinery in Houston, Texas. The refinery processes about 600,000 barrels/day of petroleum, producing a broad slate of refined products including motor gasoline, diesel, jet fuel, chemical feedstocks, and lubricants. Its operations are organized into four different sections (Fuels North, Fuels South, Specialties, and Offsites).

Having started his chemical engineering career at ExxonMobil's Baytown Refinery in 1998 as a Process Contract Engineer, Florez says that in those early years, "I did a lot of troubleshooting, which helped me to learn a lot about many of the petroleum-refining operations carried out there." Since then, he has progressed through a variety of positions in the supply chain, working on various aspects of moving the slate of products described above from the refinery throughout markets worldwide—"wherever demand arises in the global market," he says.

"That was an encouraging and rewarding assignment—to see the business side of our global refinery operations, not necessarily the chemical engineering side of things. That really expanded my view, and it was really my first foray into the business side, and it really opened my eyes to how many clean products we make, and how the demand has grown all over Europe and Africa, Asia and South America," says Florez. "It's really been amazing to see the full use of all of the refinery-derived products we make out in the marketplace."

Florez then progressed to his first supervisory assignment, leading a group of machinery engineers at the Baytown Refinery and Chemical Plant. "The company knew I did not have the specific experience needed for this role, but it was grooming and expanding my skill sets and my toolbox, and I have to admit, it was a bit scary at first being a chemical engineer who was supervising a group of machinery engineers (mostly machinists and mechanical engineers)," says Florez. "It took some time for me to make a connection and develop a good working relationship, and to optimize my value in that role," he adds.

NEXT STOP: CHICAGO

After spending his first seven years at the Baytown Refinery in Houston, Florez and his wife Valerie (another chemical engineer, who is profiled in Chapter 5) had an opportunity to move to Chicago, so he could take on the role of Operations Supervisor at ExxonMobil's Joliet Refinery. "In this role, I was not really a technical/engineering supervisor dealing with other degreed engineers," says Florez. "Rather, I was supervising a group of process technicians who had a vast amount of skills and knowledge, but did not hold an engineering degree."

Ensuring plant safety and maintaining a failsafe safety culture was an important aspect to this position. "The units I was supervising there involved the handling of highly flammable, dangerous materials—and it was my responsibility to make sure that our staff workforce did everything possible to manage and mitigate risk, every minute of every day," says Florez.

"That really keeps you up at night—thinking about all the things that could go wrong, and wanting to always stay ahead of the best-in-class safety tools, training opportunities, automated systems, and monitoring devices—to ensure personnel safety, protect the environment, and safeguard the company's assets and products." He spent three years in that role, and eventually "started to sleep better at night," he says with a laugh.

EXPANDING HIS GLOBAL PERSPECTIVE, AT HEADQUARTERS AND OVERSEAS

With his next assignment, Florez took on two roles at ExxonMobil's Fairfax, Virginia, headquarters. First, he served as the company's Global Business Advisor, and then he served as Products Optimizer, each for roughly two years.

In the role of Global Business Advisor, Florez helped to "steward the performance of all refineries we had across the world." Through that role, "I got to see how big ExxonMobil's footprint was all over the entire globe with respect to petroleum refining," he says.

Importantly, he had the opportunity to spend time at 22 of the company's roughly 30 petroleum refineries and regional headquarters offices around the world at that time. "That was an eye-opening experience on so many levels," says Florez. "The operating environments are so different in Japan, Canada, Malaysia, Singapore, and elsewhere, and these are all cultures I'd never been exposed to, so it was an immensely informative set of experiences."

Equally important, Florez notes that these two roles at the Fairfax headquarters "exposed me to very senior levels of corporate management—the types of people who are many levels above you," he says. "It's not just about being around them. It's about really seeing how these senior-level people think, gaining an understanding of why they ask certain questions, seeing their thought processes and decision-making, as it relates to how they shape our business. It was just an amazing, eye-opening opportunity."

In his role as Products Optimizer, Florez was responsible for coordinating the supply of clean products "from the gates of our refineries, through the full supply chain, and into cars, jets, barges, and lubricants."

At the end of the day we make commodity products at the refinery—gasoline and diesel—and it's the same as all of our competitors," says Florez. "So the challenge is how can we create the most value for that same product, with the same octane level and the vapor pressure as that of our competitors? It comes down to how to utilize and take advantage of our full value chain within ExxonMobil."

"My work in this role—involving such tasks as negotiating with a retail company so they source their gasoline from us to serve all of their many gas stations—was certainly a brand new experience, and it continued to get me further and further away from direct chemical engineering. It was, nonetheless, incredibly eye-opening and informative."

Eventually, Florez moved to the Netherlands, taking on the role of Technical Manager at ExxonMobil's 200,000-barrel/day Rotterdam refinery (located just outside of Amsterdam). "Now instead of just visiting the new cultures, I had to really dive in and absorb it and learn the subtle differences by living and working there," says Florez.

Interestingly, Florez recalls one incident that challenged many of his preconceived notions. "The Dutch refinery was not as 'hierarchical' as I was used to seeing, working in the US. During one meeting, a young engineer corrected a business executive who was four levels higher, and I was stunned."

"It turns out the younger engineer was right, and the senior colleagues accepted what he'd said," says Florez. "Reflecting on that exchange, I began to think about how much value we lose because people are too afraid to speak up."

HOW DID HE ARRIVE AT CHEMICAL ENGINEERING?

Florez credits his sister (who is nine years older) with blazing a trail for him to consider chemical engineering. "My parents didn't go to college, but my sister earned a PhD from the University of California at Davis, and she was already a Professor at California Polytechnic State University when I was in high school," he notes.

"She was really an eye-opener and mentor to me, especially in the STEM (science, technology, engineering, and math) field, and she was very passionate, so I was willing to look into those types of degrees," he recalls. "She'd even take me to conferences. It was really impressive to see all the possibilities in the STEM fields and so much diversity—so it is much to her credit that I even started considering chemical engineering.

"Similarly, my 11th grade Chemistry teacher (who happened to be a chemical engineer by training) was one of the most passionate and inspiring teachers I ever had," says Florez. "She was absolutely amazing—she really connected with the students, and she shared a lot of information about the opportunities and thrill of a chemical engineering career pathway."

COLLEGE WAS CHALLENGING

I found my chemical engineering major in college to be very challenging, between Fortran, principles of chemical engineering, thermodynamics... Our class started out with about 100 chemical engineering students, and we graduated something closer to 20," says Florez. "I really wanted to be successful and make my family proud, and I wanted a successful rewarding career with good earning potential."

He remembers that there were a few semesters that were harder than others. "You have to look at your internal priorities, and you must keep them straight," he says. "I joined a fraternity and was working as a bartender, and, for a period of time, I prioritized these activities above scholastics, and my grades began to drop."

"I had to reevaluate everything," says Florez. "I knew that if I prioritized things differently, the results would change, so I conscientiously began to spend more time and focus on my schoolwork—and everything else had to come in second. I was really able to turn it around once I reprioritized, started working in small study groups, and started taking advantage of mentoring that was offered by some of my professors."

"I think that, for many college kids, it's hard to think beyond the immediate gratification," he says. "Every student needs to fail a little bit because it's how you learn from that that influences your future behavior. But you also need to ask yourself if that fraternity house is really going help build your career or create opportunities or pay the bills down the road?"

MIND YOUR GPA, BUT GET INVOLVED IN ACTIVITIES

Florez underscores how important it is that students get involved in activities that really show their leadership skills. "When I'm hiring, I'd rather take someone with 3.4 GPA (as opposed to a 4.0), who has also held jobs and internships and has been engaged in extracurricular activities," he says. "To be attractive to potential employers, students must be able to show that they are engaged in society, learn how to manage multiple responsibilities, and demonstrate some leadership and work experience." ■

PERFORMANCE CHEMICALS SECTOR

In her own words
"Connecting the dots to procure process plant equipment and systems"

JILL ROGERS, DuPont Performance Materials

Current position:
Large Capital Project Team Leader, DuPont Performance Materials (Orange, TX)

Education:
• BS in Chemical Engineering, Drexel University (Philadelphia, PA)

A few career highlights:
• Led a project to replace a sulfur furnace (a 40-foot-long rotating kiln reactor) in LaPorte, Texas
• Led the effort to retrofit a refrigerants process in Corpus Christi, Texas, to make another refrigerant product; with some piping modifications and an alternative reactor, the plant could then swing from one product to another, as business needs dictated
• Led a project to install a new reactor in Orange, Texas, to meet changing capacity demand; this new reactor involved state-of-the-art metallurgy. Business expectations were exceeded following this project and capacity was even higher than predicted

Jill Rogers is Large Capital Project Team Leader for DuPont Performance Materials at the company's Sabine River Works Facility. In this capacity, she leads a team of engineering, design, procurement, and construction personnel, executing large capital projects to meet business needs.

"When a particular chemical process needs to replace or add equipment—to support capacity expansions or safety improvements, to address mechanical integrity issues, or to support the production of new products—such initiatives trigger the modification, replacement, or upgrade of equipment," explains Rogers. "The technical organization associated with the equipment does a lot of the upfront homework to define the need and consider how it should be accomplished, and then they hand the challenge over to my team so that we can fill in all the gaps."

"We determine the appropriate materials of construction and thicknesses of the needed piping and appropriate routing, identify the necessary control strategy, contemplate how is it going to be supported (in terms of civil and structural aspects)," she continues. "My team integrates the knowledge of many experts, and we start the procurement process to get the system built."

MAKING THE MOST OF HER CHEMICAL ENGINEERING EDUCATION

As a teenager, Rogers had several primary influences that opened the door for chemical engineering. "My father was an industrial/manufacturing engineer, so I had that influence in my life, and I always loved math and science, which always came easy to me. I also appreciated my high school Chemistry teacher and what I learned from her," she says. "While I enjoyed chemistry, I knew I did not want to go on to earn a PhD in Chemistry. I also knew that engineering was a good career option, even with just an undergraduate degree, so my decision to major in chemical engineering was a great way to marry the two."

Rogers encourages students to take full advantage of the office hours of professors and teaching assistants to get the time and attention needed to work through the difficult concepts. She adds: "I was never one of those people who would raise my hand during a crowded lecture to say 'Wait, I don't really understand this,' so going in for office hours was a better approach for me."

Similarly, she stresses the importance of creating smaller study groups with peers. "I constantly encouraged groups of classmates to study together," she says. "Sometimes classmates understand the concepts before you do, and they are able to choose the right words to explain a concept you may be struggling with, so you can really understand it."

A COLLEGE CO-OP OPPORTUNITY THAT LAUNCHED A CAREER

Rogers notes that Drexel had a strong co-op program that allowed all students to be in school for six months and work for six months, in their future fields, and graduate with a BS in five years. She recalls fondly the impact that that experience has had on her life. "I was fortunate during my college years to have had three different co-op assignments within a particular DuPont plant site (one involved with raw materials, one involved in finishing, and one involved in manufacturing). The experience led to a job offer upon graduation, and, 28 years later, I'm happy to say I'm still with the company," says Rogers.

"One thing I valued in my co-op work was that the school work was so theoretical, dry, and technical, but in the workplace, I was learning more practical things (such as how to size a pump) and that was what I really enjoyed doing—so that inspired me to get through the course work to be able to launch my career," she adds.

Rogers notes that such co-op programs serve a dual purpose—future employers can vet out students to see if they would be a good fit for employment following graduation, and students can get a feel for the type of work they may or may not want to pursue once they've completed a chemical engineering degree. "And, the experience itself makes you very attractive to any potential future employers, even if you don't end up working for that company," she adds.

SOMETIMES A FORK IN THE ROAD APPEARS—AND THAT'S OKAY

During her early years at DuPont, Rogers was moving up the corporate ladder, and she says: "I expected that my entire career would be in Operations Supervision, because I loved that work and always had."

But, around 2001, there was a "reduction in force" at DuPont and someone who had been executing the Capital Projects piece was let go. "My Supervisor asked me to try the role, and I've been doing that now since 2001, starting out by managing smaller projects, and eventually moving on to larger projects," she says. "While I always enjoyed my earlier role, that new opportunity opened up a new door, and I've been really enjoying my work in Capital Project Leadership ever since—perhaps even more so, and that's not something I could envisioned myself doing or predicted early on."

"It's important early in your career to try different things and to keep talking to supervisors about other types of work or roles you might want to try to broaden your experience base and build skills," she adds. "Otherwise it's really easy to become 'stove-piped' as you become more and more senior. By contrast, if you're able to move around a lot in the first 10 years or so, then you can start to streamline and focus on what types of senior role you might like best."

For example, at two points in her career (but not currently), Rogers also served as the Site Engineering Leader for two DuPont sites, managing the entire group and all functions that had to do with executing projects. "With my education and the knowledge I have acquired in both chemical engineering and mechanical engineering, as a Manager, I am able to understand the technical challenges that occur on my projects," says Rogers.

REFLECTIONS ON LIFE AS A FEMALE CHEMICAL ENGINEER

Rogers is aware that female engineers working in the traditionally male-dominated chemical process industries can sometimes encounter hostile reactions from male peers, colleagues, or supervisors. "My advice: If you feel like people are treating you differently, then make sure your supervisors are aware."

"I have been very fortunate to have worked for a company that has always fostered a very open and equal atmosphere toward all engineers and I have never really felt that people viewed me or treated me as being different or inferior because I am a female engineer," says Rogers. "I have also been very fortunate to have held important, high-profile roles that had already been charted in the past by women, so I never felt like I was charting new ground."

"Knock on wood, this is where I'll retire," she says, with a smile. ∎

REFERENCES

AIChE. 2008. A Century of Triumphs: Ten Lasting Chemical Engineering Achievements. *Chemical Engineering Progress*, November, 5–25.

American Institute of Chemical Engineers. 1992. *Ten Greatest Achievements of Chemical Engineering*. New York, NY: American Institute of Chemical Engineers.

Solen, Kenneth A. and John N. Harb. 2011. *Introduction to Chemical Engineering*, 5th Edition. Hoboken, NJ: John Wiley & Sons, Inc.

US National Academy of Engineering. 2017. www.engineeringchallenges.org/challenges.aspx and www.nae .edu/MediaRoom/20095/164396/171178.aspx. Accessed November 12, 2017.

Historical Adventures in Chemical and Biomolecular Engineering

2

"Man is a singular creature. He has a set of gifts, which make him unique among the animals: so that, unlike them, he is not a figure in the landscape—he is a shaper of the landscape. In body and in mind he is the explorer of nature, the ubiquitous animal, who did not find but has made his home in every continent."

Jacob Bronowski, *The Ascent of Man* (1973)

This chapter reviews many important inventions by applied scientists and chemical or biomolecular engineers. This discussion aims to highlight the breadth and depth of contributions that such technical professionals routinely make to society.

THE PRODUCTION OF SAFE DRINKING WATER

Many parts of the world are desperate for an adequate supply of fresh water that is safe to drink. Practical sources to consider in the production of potable water include seawater, brackish water, and purified wastewater streams, but each of these brings its own technical challenges. The chemical engineering unit operation called distillation (see Chapter 6 for a description of distillation) has been used to purify water streams to make them potable, but this approach consumes a lot of energy and, due to its technical complexity, distillation is not suitable for use in all regions of the world, particularly in developing nations.

In 1960, Sidney Loeb and Srinivasa Sourirajan, two chemical engineers at the University of California, Los Angeles (UCLA), created an anisotropic, semipermeable membrane made of cellulose acetate. Thanks to the design of these novel membranes, water is able to pass easily through the them, but salts that make the water unsuitable for drinking cannot (Loeb and Sourirajan 1960). The *anisotropy* (defined below) results from the use of a moderately porous layer of cellulose acetate to provide physical strength, which is overlain on the saltwater side with a thin, dense layer of cellulose acetate. This configuration provides for preferential selectivity for water over the salt that is dissolved in the water as the stream passes through the membrane. This groundbreaking invention allows potable water to be produced from seawater simply by pumping it through the membrane, separating the salts and leaving them behind.*

Anisotropy is the property of being directionally dependent, which implies different properties in different directions, as opposed to isotropy, which means that a given property is the same in all directions. Anisotropy can be defined as a difference—when measured along different axes—in a material's physical or mechanical properties (for example, absorbance, refractive index, conductivity, tensile strength, and others).

---

* Earlier versions of semipermeable membranes that were available before Loeb's and Sourirajan's invention did not offer sufficiently high permeability to water with efficient salt rejection; therefore, they could not be used on a large scale to produce potable water from seawater and other brackish sources.

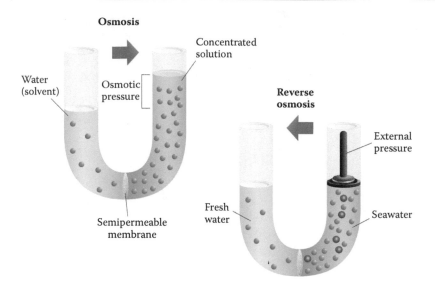

FIGURE 2.1   Osmosis and reverse osmosis are two important concepts in chemical engineering; both are used to separate dissolved solutes from a solvent (typically water or another fluid). Both processes rely on a semipermeable membrane (meaning the membrane allows certain components of the solution to pass through it, but not other components). During osmosis, the solvent (fluid) molecules are driven through the semipermeable membrane, typically toward the fluid with a higher concentration of dissolved solute. This helps to equalize the concentration of solute in solvent on both sides of the membrane. In reverse osmosis, pressure is applied to the fluid on one side to drive it across the semipermeable membrane so that the solute is retained on the pressurized side of the membrane while the pure solvent passes through to the other side. The membrane can be tailored (in terms of its pore size and construction) to be selective about which molecules or ions can pass through. (Source: Shutterstock ID 178188626)

The process that uses an anisotropic membrane to separate unwanted pollutants from liquids is called *reverse osmosis* because water flows in the opposite direction across the membrane than it does in a conventional *osmosis* process (Figure 2.1).*

Seawater has an osmotic pressure that is 30 times atmospheric pressure (30 bar, or about 440 pounds/in², or psi). Thus, to produce potable water from seawater by reverse osmosis, the pressure on the seawater side of the membrane must be greater than 440 psi. This helps to overcome the pressure differential and drive the flow of water across the membrane.

Higher pressures produce higher flux rates of water through the membrane, so the operating pressures that are typically used in reverse osmosis to produce seawater purification are typically 55 to 85 bar (800–1,200 psi). Figure 2.2 is a diagram of a typical reverse osmosis unit that uses cylindrical membranes. Figure 2.3 is a photo of an actual reverse osmosis unit that employs cylindrical membranes.

Pumping large volumes of seawater to achieve the high pressures required by a reverse osmosis system calls for a substantial amount of energy. But the energy requirement is still much lower than that required by desalination systems that rely on either distillation or another option, such as simple evaporation. Thus, reverse osmosis systems remain a viable option for regions that must consider using seawater as a source of potable water for their citizens.

As a result, reverse osmosis is now widely used for water purification. Similarly, semipermeable membranes that allow other solvents to selectively pass through the membrane while preventing the flow of dissolved solutes across the membrane show great promise for a variety of new industrial applications.

---

* Osmosis was discovered in the nineteenth century by the French physiologist Henri Dutrochet (Encyclopaedia Britannica 2017). With osmosis, if purified water is placed on one side of a semipermeable membrane and saltwater is placed on the other, water will flow spontaneously through the membrane from the purified water side to the saltwater side by a process called osmosis. The flow of water will continue across the membrane until the concentration of salts becomes almost equal on both sides of the membrane (a process called seeking equilibrium). If enough pressure is applied to the saltwater side of the membrane (known as osmotic pressure), the process of osmosis will stop. More highly concentrated saltwater has a higher osmotic pressure compared to solutions that have lower concentrations of dissolved salts in them.

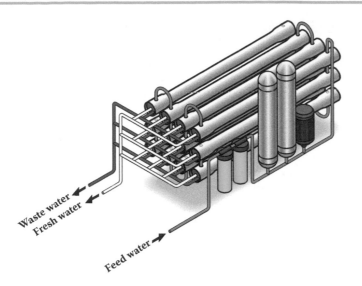

FIGURE 2.2   This diagram shows a typical reverse osmosis water-purification unit. Part of the high-pressure inlet feed water passes through these cylindrical elements, which are packed with semipermanent membranes, and the treated stream exits the system as purified water. Unwanted pollutants that were dissolved or suspended in the inlet water are retained by the semipermeable reverse osmosis membranes, and the remainder of the feedwater stream exits the facility as concentrated wastewater. (Source: Shutterstock ID 122830561)

FIGURE 2.3   This photo shows a typical example of reverse osmosis equipment that is used to produce clean water from seawater, brackish water, and other non-pure water sources. In some cases, water is treated to make it suitable for discharge to public waterways (in terms of local and federal requirements limiting the presence of specific pollutants). In other cases, polluted water streams can be treated to produce water that has been purified to sufficient levels to allow it to be reused in the chemical process facility. (Source: Shutterstock ID 500150887)

EXPLOITING THE MEDICINAL VALUE OF PENICILLIN

Penicillin was the first antibiotic product discovered and produced commercially. At the time it first became available in the 1940s, penicillin was considered to be a wonder drug (Elder 1967). Prior to antibiotics, certain bacterial infections were commonly seen as a death sentence for the afflicted. During World War II, the widespread production and availability of penicillin saved the lives of thousands of Allied service men and women.

The powerful antibacterial action of penicillin was discovered in 1928 by the British bacteriologist Sir Alexander Fleming, when he observed that the bacterial *Staphylococcus* colonies that had grown on culture plates were destroyed near an accidental mold contaminant (Figure 2.4). To develop a practical industrial process for the commercial-scale production of penicillin, it took a government-led project similar to the Manhattan Project (that had led to the creation of the atomic bomb) in scale and organization (Elder 1967).

However, penicillin was not widely available until 1944. This is because the penicillin molecule is easily inactivated, making the manufacture and isolation of commercial-scale quantities very difficult (Elder 1967). Equally importantly, the concentration of penicillin produced by the initial strain of the mold *Penicillium notatum* was too low to be a practical source of the antibiotic.

Over a decade, these two obstacles to reliable penicillin production slowed the progress of research on this promising new antibacterial compound. In fact, penicillin "nearly died on the vine," as the saying goes (Elder 1967). Had it not been for World War II, penicillin might not have been developed. It was the search for better medicines for the burns that countless victims suffered in the fire-bombing of London that led Oxford University faculty members Howard Florey and Ernest Chain to begin research on penicillin (Elder 1967). Once enough penicillin had been produced for preliminary clinical use, its life-saving potential was recognized.

To bring this life-saving drug to society, Florey and Chain enlisted the help of the US pharmaceutical industry and the US Department of Agriculture. The compelling benefits led to the creation of a coordinated government-led program to improve penicillin yields from fermentation by searching the world for strains of *Penicillium* mold that could produce high concentrations of penicillin.

Equally active research on effective methods of isolation and purification of penicillin also led to great results. In the near term, the government built several production facilities, each of which had hundreds of one-liter bottles (Figure 2.5) for growing the desired mold. After mold

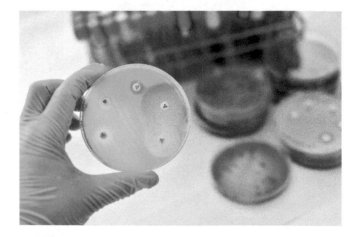

FIGURE 2.4   The antibacterial properties of *Penicillium* mold can be seen in this petri dish, which shows several zones where bacterial growth has been cleared thanks to the presence of the mold nearby. This accidental observation in a laboratory in 1928 by Sir Arthur Fleming led to the development of the antibiotic medication penicillin (excreted by the mold), which was one of the most important early breakthroughs in medicine. Since its first use in human patients in 1942, a class of antibiotics based on penicillin has provided an essential option for killing bacterial infections in patients, saving countless lives. (Source: Shutterstock ID 648642895)

FIGURE 2.5   An early penicillin bottle plant, circa 1942, when widescale production had begun to make this lifesaving antibiotic medication available to doctors everywhere for the treatment of bacterial infections. (Source: Shutterstock ID 249573532)

growth and penicillin formation was complete, the bottles were emptied, and the penicillin was extracted from the fermented liquid contained within. This involved a very labor-intensive process of filling and emptying the bottles and providing filtered air for each bottle during fermentation.

Eventually, it became practical to grow the mold in a submerged culture in fermentation reactors with multi-thousand-gallon capacity (Figure 2.6). This approach allowed for a much more economical and much less labor-intensive method than the so-called early "bottle plants."

By 1944, this major project had helped to spur the development of a pharmaceutical industry that was producing high-purity penicillin in sufficient quantities to cure millions of people from life-threatening infections. For an excellent history written by many of the participants, see Elder (1967).

REFINING SUGAR

Norbert Rillieux (1806–1894), widely recognized as one of the first chemical engineers, revolutionized the nineteenth century's energy-intensive sugar-processing industry with his invention of the first multi-effect vacuum evaporator. He (shown in Figure 2.7) is viewed by many as being one of the architects of the modern sugar industry.

Rillieux's great scientific achievement was his recognition that at reduced pressure the repeated use of latent heat would result in the production of better quality sugar at lower cost. One of the great early innovations in chemical engineering, Rillieux's invention is widely recognized as the best method for lowering the temperature of all industrial evaporation and for saving large quantities of fuel (American Chemical Society 2017).

Today, multi-effect vacuum evaporators are widely used in many industries to greatly reduce energy consumption (Figure 2.8). The multi-effect vacuum evaporator also operates at lower

FIGURE 2.6   Today, industrial-scale fermentation units, such as the one shown here, are used for the commercial-scale production of penicillin. (Source: Shutterstock ID 568754662)

FIGURE 2.7   Norbert Rillieux was a pioneering chemical engineer credited with developing the process of multi-effect vacuum evaporation (among other things), which helped to revolutionize the process of commercial-scale sugar production. (Source: American Chemical Society 2017; https://www.acs.org/content/acs/en/education/whatischemistry/landmarks/norbertrillieux.html)

FIGURE 2.8 Photo of a multi-effect vacuum evaporator in a sugar mill, which helped to revolution-ize the energy-intensive process of producing refined sugar from sugar cane and other raw materi-als, allowing the process to be carried out at lower temperatures, and thus at lower cost. (Source: Shutterstock ID 66461854)

temperatures than evaporators that do not operate under vacuum conditions. Operation at lower temperatures also helps to reduce the potential thermal degradation of product, and helps to minimize related losses and contamination of the final product.

SYNTHESIZING AMMONIA AT AN INDUSTRIAL SCALE

The world of chemical engineering is peppered with many important building-block chemicals, and ammonia is one of them. During the latter part of the nineteenth century, there was an increasing demand for nitrogen-based fertilizers that were needed to promote the crop growth necessary to feed a growing population.

At the same time, there was increasing demand for nitrogen-containing feedstocks to sup-ply the burgeoning chemical industry. At that time, mining of niter (sodium nitrate) deposits was typically the primary source of nitrogen for these two critical products. Several industrial processes were developed during the nineteenth century to synthesize ammonia (NH_3) from air and hydrogen. Ammonia is a key feedstock that is needed to produce both nitrogen fertilizers and nitrogen-based chemicals—but these early processes were inefficient, costly, and hazardous.

Ammonia (NH_3) is produced by combining molecular hydrogen (H_2) and molecular nitrogen (N_2), via the following reaction:

(2.1) $$N_2 + 3H_2 \rightarrow 2NH_3$$

This seemingly simple reaction is very challenging to carry out because the triple bond in molec-ular nitrogen is difficult to break, and because the reaction tends to be extremely slow at ambient temperatures. An enormous breakthrough occurred when the chemistry and chemical engineering communities discovered a catalyst that could accelerate the reaction; this breakthrough won Nobel Prizes in Chemistry for Fritz Haber (Figure 2.9) in 1918 and for Carl Bosch (Figure 2.10) in 1931.

Another initial limiting factor was that the reaction equilibrium of this reversible reaction shifts from being "favorable to the production of ammonia at low temperatures" to "unfavorable for

FIGURE 2.9 German chemist Fritz Haber pioneered several critical chemical engineering break-throughs in the early 20th century that allowed for the catalytic synthesis and commercial-scale production of ammonia, an important "building block" chemical that is widely used in many industries (for example, in the production of nitrogen fertilizers). Haber-Bosch discovery of the catalyst needed to produce ammonia won the Nobel Prize in Chemistry in 1918; the so-called Haber–Bosch catalytic process that allows for the production of ammonia at a commercial scale won the Nobel Prize in Chemistry in 1932.

FIGURE 2.10 Carl Bosch, who trained as a chemist but functioned as a chemical engineer, shared the 1931 Nobel prize in Chemistry with Friedrich Bergius for their breakthrough work in inventing methods for the high-pressure production of chemicals, such as the Haber–Bosch process for producing ammonia. The process is still widely used today.

ammonia production at high temperatures." Thus, increasing the temperature to the point where the reaction is fast enough to be useful also leads to relatively low conversion, according to Equation 2.1. Efforts to improve the viability of this equation eventually revealed that the use of high temperatures—typically between 400 and 500°C (752–932°F)—in the presence of a catalyst makes the reaction occur fast enough to be practical.*

The reaction is made even more favorable when it is carried out under high pressure—to the order of 5–25 MPa (150–250 bar; 2,200–3,600 psi). In Equation 2.1, four molecules of reactants (gaseous hydrogen and nitrogen) are converted into two molecules of gaseous ammonia. You can imagine the process as follows: carrying out the reaction at higher pressure pushes the reaction from the larger volume occupied by hydrogen and nitrogen molecules to the smaller volume occupied by ammonia molecules.

At the beginning of the twentieth century, many scientists were searching for ways to make ammonia. In 1909, the chemist Fritz Haber and his assistant Robert Le Rossignol demonstrated a successful catalytic ammonia synthesis process. They carried out the reaction shown in Equation 2.1 at high temperature and high pressure in the laboratory. The laboratory-scale equipment achieved a flowrate of only 125 mL per hour, but signaled a breakthrough in this pursuit.

Haber initially used osmium as the catalyst, but this was only available in small quantities. Next, he demonstrated the use of uranium as a catalyst; while more readily available than osmium, uranium also had many disadvantages.

The German chemical company BASF—sensing that this novel process showed great promise—eventually purchased the technology and assigned Carl Bosch to develop the methodologies needed to produce ammonia at larger (industrial-scale) quantities through a commercially viable process. In 1909, Alwin Mittasch, under Bosch's direction, discovered an iron oxide catalyst that proved useful in driving the ammonia reaction more effectively. Ammonia was first produced by the resulting Haber–Bosch process in the BASF plant in Oppau, Germany in 1913, eventually increasing to a commercial-scale production rate of 20 tons per day by 1914.

In the twentieth century, ammonia became the primary feedstock used to produce the three most widely used synthetic fertilizers—anhydrous ammonia, ammonium nitrate, and urea.

It is interesting to note that in 1913, when Haber and Bosch first succeeded in producing ammonia on an industrial scale (Figure 2.11), the population of the world was about 1.75 billion. Since then, the world's population has grown to 7.5 billion, underscoring the critical role that fertilizers play in ensuring that sufficient volumes of food can be produced from planted crops. Today, it is estimated that 60 percent of the earth's land would be required to feed the world's population using nineteenth-century agricultural methods. However, thanks to the widespread availability of synthetic fertilizers and other agrichemicals, as well as modern agricultural practice, only 15 percent of the earth's land is currently required to produce the crops needed to feed the world.

As recognition of the critical importance of his discovery, Haber received the Nobel prize in Chemistry in 1918. For his successful scaleup of this high-pressure process and for the development of a further improved catalyst, Bosch received the Nobel prize in Chemistry in 1931.

It is worth noting that both Haber and Bosch were trained as chemists. However, because of their focus on developing the engineering methodologies needed to address the challenges of optimizing their promising ammonia reaction at industrial scale—and managing the challenges of operating the process safely and effectively at both high temperatures and high pressures—it is fair to say that both Haber and Bosch functioned as early chemical engineers.

PRODUCING ACETONE FROM FERMENTATION

Fermentation is the process of using a microorganism, such as yeast, fungus, or bacteria, to produce a desired product, such as alcohol. Microorganisms can produce many different products at

---

* A catalyst is a substance that promotes or accelerates, but is not consumed by, a desired chemical reaction.

FIGURE 2.11   Today, world-class ammonia plants, like the one shown here, still use the Haber–Bosch process to produce ammonia from nitrogen and hydrogen. (Source: Shutterstock ID 375891859)

ASIDE: THE NITROGEN CYCLE

Today's chemical and biomolecular engineers must be aware of the potential environmental impacts of industrial processes so that ways to minimize their impacts can be devised and employed. Engineered processes that convert nitrogen to ammonia have now become a large part of the global *nitrogen cycle*. (Nitrogen is an essential ingredient in every living thing in the biosphere; the nitrogen cycle is the name of the physical, chemical, and biological processing of nitrogen-containing components in nature.) Man-made activities affecting the nitrogen cycle can have adverse consequences. For example, the leaching of nitrogen fertilizers from farms along the Mississippi River watershed allows elevated levels of these fertilizers to flow into the Gulf of Mexico. This has led to the formation of dead zones—some as large as 8,400 square miles—where water contains no available dissolved oxygen, resulting in the death of aquatic plants and fish. Improved agricultural practices can help to minimize the environmental impacts of using nitrogen fertilizers to promote crop growth by reducing the leaching of fertilizer from land to nearby waterways. Chemical and biomolecular engineers routinely develop ongoing solutions to help minimize the harmful effects of human activities on the global carbon and nitrogen cycles.

mild temperatures and atmospheric pressure, making them attractive to biomolecular engineers to explore and exploit as agents of industrial (bio)chemical change.

In 1916, Chaim Weizmann (Figure 2.12) isolated a strictly anaerobic bacterium *Clostridium acetobutylicum** that could produce acetone, butanol, and ethanol from sugar via a process called ABE fermentation (Thiemann 1963; Jones 1986). Acetone has many uses, both as a water-miscible solvent and as a chemical feedstock. During World War I, the British used acetone as a key ingredient in the production of cordite, a family of smokeless propellants that was widely used as a replacement for gunpowder (Jones 1986).

Weizmann has been called "the father of biomolecular engineering" for his pioneering work on ABE fermentation methodology. In the mid-twentieth century, the largest industrial plant to use ABE fermentation to produce acetone was in Peoria, Illinois. The plant carried out the

---

* Anaerobic microorganisms thrive without the need for oxygen.

FIGURE 2.12   Chaim Weizmann, an Israeli chemical engineer who developed a fermentation process to produce acetone. Fermentation, like many other biological processes that rely on microorganisms, allows many chemical conversions to be carried out at milder temperatures and pressures compared to conventional chemical process operations. Today, biomolecular engineers are in great demand to help explore and expand the many ways in which biological processes can be used to produce chemicals, pharmaceuticals, foodstuffs, and other products. Later in his life, Weizmann became the first President of Israel. (Source: Shutterstock ID 42909280)

fermentation of molasses to make acetone, butanol, and ethanol using 96 fermenters, each with a capacity of 50,000 gallons (Jones 1986).

In the 1950s, petroleum became a more economic feedstock to use for the production of acetone, butanol, and ethanol. In recent years, however, biomolecular engineers have been working to develop new technology to enable the large-scale, economic microbial conversion of biomass to produce acetone, ethanol, and related products (Figure 2.13).

FIGURE 2.13   Shown here is an aerial view of an industrial solvents plant, such as those used to produce solvents like acetone, butanol, and ethanol. (Source: Shutterstock ID 285764054)

ASIDE: THE ABE FERMENTATION PROCESS

Biomolecular engineers use microorganisms and enzymes to produce useful (bio)chemicals; the ABE fermentation process is just one example. However, biochemistry—the chemistry of life—is complicated. Living organisms carry out most biochemical reactions inside living cells that grow under ambient pressure and temperature conditions. To illustrate this inherent complexity, Figure 2.14 shows the two intracellular biochemical pathways of the bacterium *Clostridium acetobutylicum*, which is used to produce a variety of chemicals when grown anaerobically (that is, using bacteria that can thrive without oxygen) with glucose (a sugar, the primary component of starch) as a feedstock under either acidic or neutral conditions (Thiemann 1963; Jones 1986).

The ABE fermentation reaction pathway is carried out under neutral conditions (shown on the right side of Figure 2.14). Biochemical pathways like these are often complex. However, one attractive factor is that such conversion reactions, which are catalyzed by enzyme activity, are typically carried out at ambient conditions (meaning there is no need for elevated temperatures or pressures), which typically simplifies the equipment and safety systems needed to carry out the process on a commercial scale. By comparison, many chemical conversions that are used to produce a given product often require higher temperatures and pressures (for example, the synthesis of ammonia, as described above). These higher temperatures and pressures often use special process equipment that is expensive to make, operate, and maintain—possibly creating additional safety and environmental risks.

A new direction in biomolecular engineering is focused on creating new biochemical pathways that are catalyzed by modified enzymes using genetically modified organisms (GMOs). GMOs have the potential to provide great benefits to society. Since prehistoric times, people have sought to improve agricultural yields by selective plant breeding. In the twentieth century, mutagens were used to create mutant microorganisms that produced higher concentrations of antibiotics and other useful compounds. In these prior efforts to create improved strains of living organisms, the changes in the genetic material of the organisms were unknown. The resultant new organisms were simply substituted for the previous organisms if they offered improved properties; the makeup of the genes was randomly changed.

Today, with GMOs, conscious and specific changes are made in the genetic makeup of the native organisms to produce the specific desired improvements. No longer accidental, these genetic changes can be directed to yield the desired result.

However, as with any new and powerful technology that has the potential to be misused or abused, society needs to have methods in place—as well as healthy intellectual and ethical exploration and debate—to ensure that GMOs do not produce undesirable effects. For example, domestic plants that have been bred to be resistant to herbicides might transfer that gene to wild weeds that had previously been controlled with that herbicide.

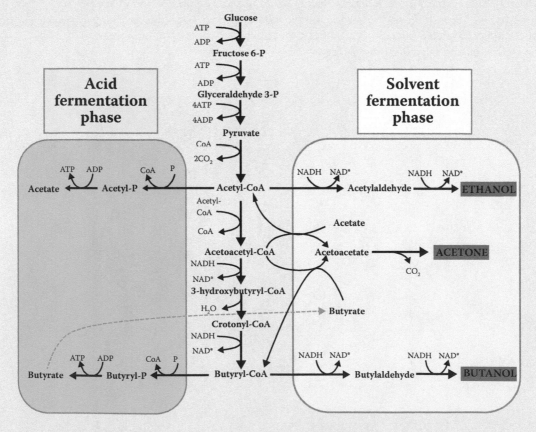

FIGURE 2.14   Two alternate biochemical pathways of *Clostridium acetobutylicum*. These biochemical pathways are shown here to illustrate the complexity of the microbial biochemical steps to convert glucose (dextrose) into industrial solvents. (From Jones, DT., *Microbiological Reviews*, 50, No. 4, 484–524, 1986.)

REFERENCES

American Chemical Society. 2017. National Historic Chemical Landmarks. Norbert Rillieux and a Revolution in Sugar Processing. www.acs.org/content/acs/en/education/whatischemistry/landmarks /norbertrillieux.html (accessed May 21, 2017).

Elder, Albert L., ed. 1970. The History of Penicillin Production. *Chemical Engineering Progress Symposium Series*, 66, No. 100, 100 pages. New York: American Institute of Chemical Engineers.

Encyclopaedia Britannica. 2017. Henri Dutrochet, French Physiologist. www.britannica.com/biography /Henri-Dutrochet (accessed June 2, 2017).

Jones, David T. 1986. Acetone-Butanol Fermentation Revisited. *Microbiological Reviews*, 50, No. 4, 484–524.

Loeb, Sidney. 1966. *Desalination*, 1, 35; with Srinivasa Sourirajan, 1960. Report 60-60, Department of Engineering, UCLA, Los Angeles, CA.

New York Times. 1952. On This Day: Chaim Weizmann Of Israel Is Dead. www.nytimes.com/learning/general /onthisday/bday/1127.html (accessed June 15, 2017).

The Nobel Prize in Chemistry in 1918. www.nobelprize.org/nobel_prizes/chemistry/laureates/1918/ (accessed June 14, 2017).

The Nobel Prize in Chemistry in 1931. www.nobelprize.org/nobel_prizes/chemistry/laureates/1931/bosch -facts.html (accessed June 14, 2017).

Thiemann, Kenneth V. 1963. *The Life of Bacteria*, Second Edition. New York: Macmillan Company.

Where Do Most Chemical and Biomolecular Engineers Work? How Much Are They Paid?

3

Due to the rigor of their undergraduate and graduate school education, graduates who hold one or more degrees in chemical engineering and biomolecular engineering tend to enjoy good starting salaries and strong opportunities for career and salary growth throughout their varied careers. This chapter briefly summarizes recent US statistics about where chemical and biomolecular engineers work and how much they are paid.

Chemical and biomolecular engineers work in a very diverse array of industrial sectors, including but not limited to:

- Many types of manufacturing
- Pharmaceuticals discovery and production
- Healthcare
- Engineering design and construction
- Pulp and paper production
- Petrochemicals production
- Food processing
- Specialty chemicals production
- Microelectronics
- Electronic and advanced materials
- Polymers production
- Business services
- Biotechnology
- Environmental, health, and safety (EHS) industries
- Law
- Law enforcement
- Public service

Of the 34,300 jobs classified as "chemical engineering" by the US government in the latest (2014) data (see Table 3.1), median annual pay for those with a chemical engineering degree was US$98,340 per year, or about US$47 per hour (US Bureau of Labor Statistics 2017). Average salaries for chemical engineering positions in the US typically started from US$59,470 and went up to US$157,160, according to the same US Government data source (US Bureau of Labor Statistics 2017).

Similarly, in a recent *Chemical Processing* magazine survey (Joshi 2017) of 875 employees in the chemical process industries, the magazine reported an average salary of US$112,637. Note that many of the magazine's readers are chemical engineers, but chemists and mechanical engineers may also represent some of the respondents to the magazine's survey.

It is worth noting that in the experienced opinion of these two authors, the US Department of Labor wage statistics do not truly reflect the full salary potential of the large number of chemical and biomolecular engineers who have been promoted or transferred to many other burgeoning and non-traditional careers and job titles that are discussed in this book (see Table 3.2 for examples of still more career pathways for chemical engineers and biomolecular engineers who are not otherwise described in this book). In our experience, chemical and biomolecular engineering

TABLE 3.1   Industry Sectors that Provide the Largest Employment of People in the US Who Are Classified as Chemical Engineers

• Engineering services	16%
• Basic chemical manufacturing	14%
• Research and development in the physical, engineering, and life sciences	10%
• Petroleum and coal products manufacturing	7%
• Resin, synthetic rubber, and artificial synthetic fibers and filaments manufacturing	7%
• Other	46%

Source: US Bureau of Labor Statistics (BLS), Federal statistics on chemical engineers, www.bls.gov/ooh /archi tecture-and-engineering/chemical-engineers.htm, (accessed on June 1, 2017), 2017.

TABLE 3.2   Examples of Other "Non-Traditional" Jobs Where a Chemical Engineering Degree Is Useful

- Analytical chemist
- Energy manager
- Manufacturing engineer
- Materials engineer
- Mining engineer
- Production manager
- Quality manager

Source: US Bureau of Labor Statistics (BLS), Federal statistics on chemical engineers, www.bls.gov/ooh/archi tecture-and-engineering/chemical-engineers.htm, (accessed on June 1, 2017), 2017.

graduates can obtain even higher salaries than those cited above when they ascend the technical and managerial/executive ranks in their chosen professions.

The American Institute of Chemical Engineers* (AIChE) has more than 53,000 members, including members from 110 countries worldwide. Although it is a very useful and worthwhile organization that supports the chemical engineering profession, not all chemical and biomolecular engineers are members of AIChE. Even so, the median salary (*Chemical Engineering Progress* 2017) of chemical engineers surveyed and reported by the American Institute of Chemical Engineers in 2017 was US$155,000—significantly higher than the average salaries indicated by the Department of Labor and the *Chemical Processing* magazine salary surveys. Figure 3.1 symbolizes a chemical or biomolecular engineer being paid for their professional work.

Meanwhile, Dr. Thomas K. Swift, Chief Economist & Managing Director, Economics & Statistics, American Chemistry Council† forecasts that employment in the US chemical industry alone will grow from 811,000 in 2016 to 833,000 by 2018 (Swift 2017). Those figures include chemists, mechanical engineers, manufacturing and administrative personnel, and many other professionals working in the chemical industry—but chemical engineers will be among the most influential. Table 3.3 also shows that more than 5 million US jobs involve chemistry in related industries and commerce (American Chemistry Council 2017).

The pharmaceutical and biopharma industries employ many chemical and biomolecular engineers, in a diverse array of technical and managerial positions (American Chemistry Council 2017). For example, the world's largest pharmaceutical company, Johnson & Johnson, recently reported having 127,100 employees. Even the tenth-largest company, Gilead Sciences, recently reported having 55,000 employees. Individuals holding chemical and biomolecular engineering degrees play many essential roles in pharmaceutical discovery and in the development of sustainable processes to produce life-saving drug therapies at commercial-scale amounts, and the employment potential for students graduating with chemical and biomolecular degrees is certain to remain strong, given the essential role these technical experts play in so many different industries and in so many different professional roles.

---

* The American Institute of Chemical Engineers website is available at www.aiche.org.
† The American Chemistry Council website is available at www.americanchemistry.com.

FIGURE 3.1   As discussed throughout this book (and demonstrated in the 25 profiles of working chemical and biomolecular engineers), there are many interesting, lucrative career pathways for students graduating within these two majors. According to a 2017 report from the US Bureau of Labor Statistics, the median annual pay for those with a chemical engineering degree in 2014 (latest data available at that time) was nearly US$100,000/year. (Source: Shutterstock ID 311043284)

TABLE 3.3   Total Jobs Supported by the Business of Chemistry (in Thousands), 2016

Industry	Direct	Supply Chain	Payroll-Induced	Total
Chemicals	811			811
Agriculture		73	56	129
Mining		80	11	91
Utilities		36	10	46
Construction		54	29	83
Manufacturing, excluding chemicals		475	130	605
Wholesale trade		285	85	370
Retail trade		153	353	505
Transportation		191	86	277
Information		64	50	115
Finance and insurance		179	336	515
Professional, scientific, and technical services		288	141	429
Management of companies and enterprises		360	29	389
Administrative and waste management services		266	165	430
Healthcare and education		1	597	598
Arts, entertainment, and recreation		29	88	118
Accommodation and food services		62	325	387
Other services		68	253	321
Government		45	35	79
Total: all industries	811	2,707	2,779	6,298

Source: US Bureau of Labor Statistics (BLS), Federal statistics on chemical engineers, www.bls.gov/ooh/architecture-and -engineering/chemical-engineers.htm, (accessed on June 1, 2017), 2017; "2017 AIChE Salary Survey," Chemical Engineering Progress, pp. 14–23, June 2017, American Chemistry Council, www.americanchemistry.com/Jobs/ (accessed June 15, 2017).

Note: Since 2012, the calculation of the job multipliers has been based on analysis using the IMPLAN model. The IMPLAN model is a widely accepted and robust input–output model. It calculates industry multipliers using complex output-labor ratios and industry spending patterns. Multipliers generated by this type of model are typically used in assessing the economic impact of projects or plant closures, import substitution analysis, centralized planning, and economic benefit analyses. As a result, the figures presented above are not comparable with data published in earlier editions (pre-2012) of the *Guide to the Business of Chemistry*.

INDUSTRY VOICES

"I develop and improve editorial materials that are aimed at helping chemical engineers to do their jobs better and understand the new technologies and business trends. I'm always handling a variety of topics related to both established and new topics, in an array of areas, so the work never gets dull."

—Mark Rosenzweig,
Chemical Processing
magazine,
page 39

CHEMICAL ENGINEERING—PROFESSIONAL DEVELOPMENT

In her own words
Helping to shape the entire
"lively and invigorating" chemical engineering profession

**KRISTINE CHIN, AMERICAN INSTITUTE OF CHEMICAL ENGINEERS
(AIChE; NEW YORK, NY)**

Current position:
Director of Conferences, American Institute of Chemical Engineers (AIChE)

Education:
- Bachelor's degree, Chemical Engineering (The Cooper Union, New York, NY)
- MS degree, Chemical Engineering, with a minor in Biomedical Engineering (Cooper Union, New York, NY)

A few career highlights:
- The launch of *ePlant* magazine, as a new, start-up sister publication to *Chemical Engineering* magazine
- The youngest person to serve as Editor-in-Chief of *Chemical Engineering Progress* (*CEP*) magazine; during her tenure, she helped to refresh and overhaul its editorial content (for instance, adding regular monthly news features to its existing engineering articles) and turned its annual fiscal budget from negative to positive
- Has enjoyed the honor of working with so many dedicated scientists and engineers to consistently create the best possible AIChE conference and related events

Kristine Chin has spent most of her life working for the American Institute of Chemical Engineers (AIChE), in a variety of professional roles, most recently as Director of Conferences. She earned both a Bachelor's degree and a Master's degree in Chemical Engineering (with a minor in Bioengineering) from Cooper Union in New York City.

Upon graduation, Chin took a laboratory-based position working for a manufacturer of filtration components and systems, testing filter products and systems that were aimed at electronics applications, to evaluate particulate-breakthrough thresholds. She recalls being decked out in full protective gear to guard against contact with hydrochloric and hydrofluoric acid, and realizing that she did not love the work—not only because it involved high risk, but because she felt it was also very isolating. She saw the same few co-workers every day, in a lab that had no windows, and she knew that this was not the career path for her.

MAKING A MOVE INTO PUBLISHING

Eventually, Chin decided to try something new, and joined the editorial staff of McGraw-Hill's *Chemical Engineering* magazine, one of several flagship trade journals that serve the professionals who work throughout the global chemical process industries (CPI). Always up for a new challenge, Chin says: "I took the job as an Assistant Editor, and figured, if worst came to worst, I could always go back to 'working as a chemical engineer' in industry."

"In the beginning, I thought I'd never make it as a writer and editor!" she recalls, with a laugh. "Those early assignments... oh my gosh! The senior editors 'bled all over them with red pens.' But eventually, I realized I was surrounded by great mentors—many of the editors (all of whom had chemical engineering degrees) had been on staff for many years, and had industry expertise, too."

"I pushed myself and over time I learned to research, write, and edit much better, and I really enjoyed it," she adds.

Chin notes that working as an editor (which routinely involves interacting with other technical professionals, both on staff and at companies throughout the global CPI, visiting plants and corporate locations, and attending trade shows) is a very social career option, "and that really forced me get over my shyness." She recalls: "I remember my first few telephone interviews with sources or engineering authors whose manuscripts I was editing—my heart was pounding and my palms were sweating."

Chin spent five years at *Chemical Engineering* magazine, eventually being promoted to Associate Editor. Several years into her tenure there, she also proposed the development of a spinoff sister publication, entitled *ePlant*, which she conceptualized and launched just as the so-called "Internet economy" was starting to take shape back in the early 2000s.

During that early period, a range of Internet-based applications and methodologies began to explode onto the scene, to address (and revolutionize) all types of chemical engineering endeavors. "It was a really exciting time to be at the cutting edge of 'the next big new thing' that everyone was talking about, and to have the opportunity—as Editor-in-Chief of that new start-up magazine I had proposed—to talk to all of the key players, learn about their developments, breakthroughs, and vision, and then share it with interested readers all over the world."

Roughly a half-dozen issues of *ePlant* (as a standalone magazine) were published before it was folded back into *Chemical Engineering* magazine (to become a regular monthly editorial feature), as a result of a downturn in advertising in print magazines at that time. When *ePlant* folded, Chin decided to make a change, and she was recruited to join a competitor magazine—*Chemical Engineering Progress* (*CEP*), the flagship publication of the American Institute of Chemical Engineers (AIChE), which reaches over 50,000 members.

BREAKING A GLASS CEILING

When Chin was recruited by AIChE at the age of 29 to join *CEP* as its Editor-in-Chief, she was the youngest ever to hold that position. She recalls: "I walked in and everybody there seemed to be in their 50s, and I remember feeling very nervous. But it was a really great group of people, all top-notch and well-seasoned editors. And everyone recognized that I was able to bring some fresh perspective from the years I had spent as an Editor at *Chemical Engineering*."

In particular, she was able to make good use of her strong news-writing capabilities, which added fresh perspective to *CEP*'s heavy focus on technical/engineering articles, and her entrepreneurial and business-development experience, which had been honed during her work to conceptualize, launch, and produce *ePlant* during its lifetime.

After two years at *CEP*, in the midst of a restructuring at AIChE, Chin took on the additional role of Publisher of the magazine, too, essentially managing "both the editorial side of the magazine, and the revenue-generating, advertising-sales side of the house." She recalls: "It was definitely a challenging period in my career, but also sort of fun, because I learned so much about the sales and production side of publishing, the financials and profit-and-loss (P&L) statements, and all about the complex circulation audit (a semi-annual exercise that is carried out with a third-party group, to provide advertisers with detailed information about a magazine's readers)."

MOVING FROM MONTHLY EDITORIAL WORK TO CONFERENCE PLANNING

After juggling both the Editor-in-Chief and Publisher jobs concurrently for several years, plus giving birth to her third child, Chin felt tired and burned out from the pace, so she sought to make another change within AIChE. At that time, the Programming Manager position opened up within AIChE, and although the position was a "step down" in terms of responsibility and salary, Chin says: "It was the break I needed to regain my work-life balance. Plus, I realized that I could still make an impact in this new role. I decided to take a chance and see what I could make of it."

"I found I loved it—I absolutely loved all of the people interaction at the many AIChE conferences I was working on throughout the year," says Chin. "There's a great energy associated with developing professional conferences to help working professional get together to share their work and learn about other cutting-edge trends that can help to improve and shape their own work," she adds.

Early on, Chin focused her efforts on organizing and running many local and regional AIChE meetings and conferences, such as the annual AIChE Spring Meeting and AIChE's Process Development Symposia series—all events that focus on the "industrial" side of the CPI, where she feels more at home, given her professional experience. (By comparison, she notes that she had not really worked on AIChE's Annual meeting, which tends to focus more on showcasing research and development advances that are often more academic in nature.)

In 2011, Chin got another promotion within AIChE, when she was named Program Development Director for AIChE. With that, she expanded her universe, taking over the program development of AIChE's Annual Meeting, too. "That really opened up my eyes and my vantage point, and I got to get a better sense of the full contribution and passion that is coming from the academic side of the chemical engineering profession—something that chemical engineers working in industry often don't get to see," says Chin. "Working so closely with professors, graduate students, and undergrads when developing the conference programs for AIChE's Annual Meeting also highlights emerging areas of research that continue to expand and shape the entire realm of chemical engineering, and highlights the importance of mentoring for students who are working toward their degrees."

"During the first Spring Meeting I attended, the attendance was a little over 900; when I was promoted to Programming Director, attendance was in the 2,100+ range," she says. "Since I've been overseeing it, AIChE's Annual Meeting has grown from 4,000 to more than 5,000 attendees," she notes with pride.

"While working on conferences that showcase industrial breakthroughs, I'm always thinking 'what is the practical implication of this?'" says Chin. "But when I consider the topics and breakthroughs and presentations that were coming out of college and university settings, I realize that we are really looking at critical areas of research that will impact much longer horizons—and these activities foreshadow what the breakthroughs may be in the next 10 to 20 years," she says. "It's also a good lesson to be reminded that not all of the academic research that is developed and reported at AIChE meetings will end up being successful."

Similar to learning all aspects of publishing, from editorial to production to sales, Chin was promoted once again (in 2014) to become AIChE's Conferences Director. In that role, she holistically looked at conferences from not only the program-development perspective, but also in terms of the complex logistics and financial considerations that were required to run events involving thousands of attendees.

Most recently, Chin was promoted to become AIChE's Conferences & Education Director, adding oversight of AIChE's Education department to her list of responsibilities. "Conferences and education go hand-in-hand, with the primary focus of training young engineers or advancing the knowledge of seasoned engineers," she says.

WHY CHEMICAL ENGINEERING?

"It's hard to know what you're going to like when you're in high school," says Chin. "I thought I was going to be a doctor, but with my first class in gross anatomy, I realized I just didn't have the stomach for it," she recalls with a laugh. "I was always good at math and science, and I always loved chemistry (even physical chemistry and organic chemistry), so chemical engineering seemed to be a good way to combine all of that." When she was pursuing her Master's degree, she added a concentration in Biomedical Engineering (a burgeoning course of study during the 1990s), sensing that it would have greater relevance in the coming years.

Despite her obvious bias about the inherent importance and value of active AIChE membership and participation, Chin says: "So often, at our annual student conference for undergraduates, I see a majority of students who have no idea what they should be doing beside completing their degree."

"It's important to attend these types of industry events—Not just to meet and listen to lots of really inspiring people, but to really get the full appreciation of the incredible societal impact that chemical engineers have for everyone," she continues. "Today's chemical engineers are really intent on using their ingenuity and intellect to address all of the 'grand challenges' we face as a society—in terms of feeding the world, addressing pressing fuel challenges, developing medicines and more."

"I like to tell young engineers—You're not 'just another engineer'… you may end up being *the* engineer who really makes a difference. This should inspire you to do more than just work your job, 9 to 5, and go home and live your life in an isolated way." Rather, she encourages all chemical engineers, but particularly young ones, to find interests and passions and get together with others who share them.

"Don't be passive," she adds. "It's much more satisfying to take a chance, and to really fight for the things that are really important."

"I've been very lucky throughout my career, and in the right place at the right time to take advantage of a few pivotal changes," she notes. "I can truly say I have loved my entire career trajectory. It's been an incredible privilege to be able to keep my finger on the pulse of this lively and invigorating profession." ■

ENVIRONMENTAL MANAGEMENT AND CONTROL

In his own words
"A self-proclaimed 'jack of all trades,' thanks to my chemical engineering training and expertise"

TOM MCGOWAN, PE, TMTS Associates

Current position:
Founder and President, TMTS Associates, Inc. (Atlanta, GA)

Education:
- BS, MS, Chemical Engineering, Manhattan College (New York, NY)
- MS, Industrial Management, Georgia Tech (Atlanta, GA)

A few career highlights:
- Received the Personal Achievement Award in Chemical Engineering from *Chemical Engineering* magazine (2010), for his distinguished career in solving process challenges and enhancing the field
- Received the Distinguished Service Award at the 2010 International Thermal Treatment Technology Conference
- Awarded a US Deptartment of Energy National Energy Award (1984)
- Contributor to Perry's *Chemical Engineer's Handbook* and several related professional handbooks
- Editor/Chief Author of *Biomass and Alternate Fuel Systems: An Engineering and Economic Guide*

Tom McGowan runs a small engineering consulting firm that specializes in thermal systems and air-pollution control, bulk-solids handling, and industrial ventilation. The path that led to his current role was far from straight, but every step along the way both benefitted from and enriched his chemical engineering expertise.

"Far from being on a single track, I started out in industrial-scale Research and Development (R&D), then worked for the Georgia Tech Research Institute on energy, combustion, and gasification, and went on to work for an equipment manufacturer making exotic, high-temperature oxygen burners," says McGowan. "Then I spent a stint working in hazardous waste incineration as a consultant, and then served as a Vice President of Thermal Treatment for the remediation of hazardous waste-contaminated soils, and provided consulting services for a small firm that was later sold to a large company. Eventually, in 1998, I opened up my own consulting firm." When asked recently by graduating students at Manhattan College on how he planned his career, he said he did not—"he just wanted to keep as may doors open as possible and to close as few as possible."

McGowan notes that one set of expertise often begets another. For instance, he notes that early work he did in bulk-solids handling helped him to get involved with designing baghouses and cyclones to capture particulate matter from emissions streams, as well as pneumatic conveyors and rotary-airlock systems for a variety of chemical process operations. Similarly, he says "at some point, I got involved with clean energy, gasification systems, solar systems all of; heating, ventilation, and air control (HVAC) systems; and hazardous-waste incineration—all of which require expertise in high-temperature burners, heat transfer, condensation, and other chemical engineering principles."

"Early on, I was involved in the complete design, fabrication, installation, and testing of industrial-scale equipment, which gave me a feel for data—both good and bad," says McGowan.

"I've had a checkered past, and I'm happy and proud of that," he says with a laugh. "I've been able to use my chemical engineering experience and move with the markets." He adds that it "seems harder than ever to stay with one market or one job for an entire career given the nature of the current economy and pace of change."

Calling himself "fearless and flexible," McGowan says: "If you have a good background, then picking up a new subject is not that big a deal. If you crack the books you can become an expert pretty quickly, and if it is a new market and new subject matter, how many people out there will already know more than you do? I've worn through the bindings of many of the textbooks I accumulated during my studies at Manhattan College and Georgia Tech, and have kept up with ongoing technology advances over time." "I have many, many skills and I'm not afraid to get dirty and go inside the equipment when needed," he adds. For instance, he notes that "confined-space entry keeps you honest about what you know and don't know, and takes the guess work out of troubleshooting. It also earns the respect of both engineers and operators on job sites."

One of his most significant personal achievements was being involved in pioneering work in the thermal treatment of soils contaminated by organic contaminants. "I entered the market when high-temperature transportable incinerators were first being used and, for a time, the one that I fielded at Envirite had the highest thermal capacity of any in existence. This segued rapidly into lower-cost, lower-temperature thermal treatment with higher capacity for a wide variety of contaminates, such as PCBs, coal tars, dioxins, and pesticides." This work has taken him all over the globe, as the US market matured and other countries began focusing more on reducing the environmental impact of industrial operations.

McGowan's willingness to always listen to clients and consider whether he could address their needs led to work for NASA on multiple projects for the International Space Station and Mars Landing projects. What, at first, might have seemed out of his area of expertise turned out to involve applying industrial project knowhow to their microgravity needs for retrofit filtration, CO_2 removal, and fire-hood design and prototype testing. "I did volunteer to go up on orbit for service calls, but, sadly, they have not taken me up on it," he adds, with a smile.

NETWORKING NEVER GOES OUT OF STYLE

"When a large company pulls out of a city or downsizes in an industry, a lot of smart people are out of work, but they know a lot," says McGowan, adding: "As a consultant, it's always been helpful for me to 'accumulate people' who might be useful on future projects, to help me with networking of my own, or to make an effort to help them find their next job."

"If you work at it, your contacts will accumulate easily, and these days (with email and social media), it's easier than ever to keep in touch with people," he says. In fact, for many years, McGowan has been sending out periodic mailings or emails, and maintaining his own website, to profile his expertise and detail the technical papers he has published, and the engineering seminars and webinars he will be presenting at industry meetings or online. "Whenever I do that, people remember I'm here and often they will call or email to catch up, and then networking and project opportunities arise easily."

McGowan enjoys presenting tutorial-style seminars and webinars to share his knowledge and expertise. "It's great to be able to give back and transfer the knowledge to the next generation of chemical engineers," he says. "My goal is to keep people from making avoidable mistakes, where equipment and systems are poorly designed, or the wrong equipment is specified for the job. It's essential to keep people safe and protect equipment and environment."

Similarly, McGowan enjoys his participation in a variety of professional memberships, including the Air & Waste Management Association (AWMA), Sigma Xi (The Scientific Research Honor Society), and the American Society of Mechanical Engineers (ASME). He has also served as President of his local chapter of the American Institute of Chemical Engineers (AIChE) and the Georgia Solar Energy Society.

HOW IT ALL BEGAN

Like many other chemical engineers, McGowan showed early interest and proficiency in math and chemistry. In addition to encouragement from his high-school counselor, as a teenager, his mother used to take him and his five sisters to Manhattan College (located nearby) to attend a semi-annual event called Campus Days. "I could see what the students and professors were doing in the big and small laboratories—biology labs, hydraulics labs, civil engineering labs, and more—and it utterly captivated my attention to see all of the amazing science and engineering that was being carried out there," says McGowan. "There was literally no turning back or second-guessing my desires once I'd had that exposure."

McGowan has a strong interest and natural proclivity in mechanical engineering, and for a while he considered that as a possible major. "I realized I had to take an extra year of math to complete the mechanical engineering major, so I opted for chemical engineering instead," he says. "But so much of chemical engineering work also requires a strong understanding of the mechanical engineering aspects of the equipment. I'm a mechanic/machinist/welder by nature, and am happy to say that when I'm around, the machinery is happier and runs better," he says with a laugh. "My education has been a real blend of what I was wired for and what I've acquired—and I've used every bit of it every day."

Nonetheless, he recalls that the start of college was hard. "My first year at Manhattan College was the toughest. We engineering students were told to look to the left and look to the right and know that one-third of us would be gone by the end of the year," he says. "That first year I was working as hard as I knew how and I was still earning only Bs and Cs."

"I finally earned my first A in my sophomore year, and by the time I was focusing mainly on core engineering, the real engineering subjects, I was starting to rack up my As," he says. "I finally hit my stride and I was really in my element and maintained a very respectable overall GPA."

WORDS OF WISDOM FOR STUDENTS

"It's important for students to find a professor who is skilled in the student's field of interest—one who is willing to help and to advise the student along the way, someone who can help *open doors* for them. It helps to have someone you can turn to—someone who may not even be in the 'chain in command' who can set you straight, says McGowan."

"It is also critical for students to start developing a job history early on," he adds. "You may start out working for free through an internship or co-op opportunity, but you must build job skills—like showing up a half-hour early for your interview, and arriving ten minutes early every day for work. Make working a habit. Keep learning, move forward, and know when to move on, in search of the next interesting, expansive, fulfilling opportunity." ∎

ENGINEERING TRADE PRESS

In his own words
"Technical editing and editorial development
provide variety and intellectual satisfaction"

MARK ROSENZWEIG, *CHEMICAL PROCESSING* MAGAZINE (PUTMAN MEDIA)

Current position:
Editor-in-Chief, *Chemical Processing* magazine (Putman Media)

Education:
- BEng, Chemical Engineering, The Cooper Union (New York, NY)

A few career highlights:
- Elected Fellow of the American Institute of Chemical Engineers (AIChE)
- Winner of the Jesse H. Neal Editorial Achievement from the American Business Press
- Winner of the Gold Award for editorial excellence from the American Society of Business Press Editors
- Served on the Management Board of *Biotechnology Progress* magazine, a joint venture of the AIChE and the American Chemical Society

As Editor-in-Chief of *Chemical Processing* magazine—a monthly trade magazine aimed at chemical engineers and other technical professionals who develop process technology and design and operate plants—Mark Rosenzweig is happy to keep his finger on the pulse of the technology developments and business trends that are shaping the chemical industry. "Having a chemical engineering degree, and an entire career working in the technical trade press, I have a comfort level and reasonable perspective to see how existing and emerging trends and topics are affecting our readers," he says.

His work involves planning what articles and topics the magazine should cover, lining up experts from industry to author technical articles related to their areas of expertise, and identifying key sources to interview for news-related articles. "When I receive editorial content from experts in industry, it is rarely in a form that's ready for publication," notes Rosenzweig. "Chemical engineers are not typically wordsmiths by their training, and often they do not understand the specific focus of my readership, so I work with them to edit and revise their articles, as needed, to ensure accuracy and clarity."

A HALF-CENTURY OF EDITORIAL EXPERIENCE

Rosenzweig recently celebrated 50 years in the magazine business, having started his career in 1967 as an Editorial Assistant with *Chemical Engineering* magazine (then the flagship trade publication of the McGraw-Hill publishing company). While studying Chemical Engineering at New York City's prestigious Cooper Union, Rosenzweig served as Editor of the school newspaper. "I really liked writing and wanted to explore career options that would combine both sets of interests," he says.

"The year I graduated was a good time for chemical engineering graduates, and I had plenty of job interviews and several offers, but I decided I really liked the idea of working in an office—as opposed to a process plant or refinery environment—and I enjoyed technical writing and editing," he explains. Near the end of his senior year, he contacted the Editor-in-Chief at *Chemical Engineering* to request an interview and began work there even before graduation ceremonies took place.

His role on the editorial staff at *Chemical Engineering* proved to be a good fit, and he ended up spending two decades there, moving up the editorial ranks. Throughout his tenure at that magazine, he served as (among other things), the magazine's European Editor and Managing Editor.

Eventually, he went on to become part of the launch team at McGraw-Hill for a revolutionary (at that time) CD-ROM product called *Electronic Sweets*, serving as its Operations Manager. In total, he spent 23 years at McGraw-Hill.

In 1990, the Executive Director of the American Institute of Chemical Engineers (AIChE) decided that *Chemical Engineering Progress* (*CEP*), the official magazine of the organization, needed drastic overhauling to better meet the needs of practising engineers. Rosenzweig was hired to be the Editor-in-Chief at *CEP* and to engineer the editorial overhaul. "Earlier in its history, *CEP* was seen as 'just the magazine you got when you paid your membership dues,' but we worked to make it more relevant during my time there."

CHANGING PARADIGM

After spending 11 years at *CEP*, Rosenzweig switched magazines again, bringing his expertise to *Modern Plastics* magazine (a former McGraw-Hill title that was, at the time, owned by Chemical Week Associates), and spent two years there as Executive Editor.

In 2003, Rosenzweig was offered the role of Editor-in-Chief of *Chemical Processing* magazine, and given a similar mandate to what he'd heard when he joined *CEP*—to improve the editorial reach, depth, and breadth of the magazine's technical and business content and make it more relevant and valuable to its target audience of engineers in the chemical industry.

"*Chemical Processing* is published by a small, family-run company (Putman Media) based in Schaumburg, Illinois, and when I was hired they had never had an Editor-in-Chief work remotely," says Rosenzweig, "but my family was settled in New York and I didn't want to relocate to suburban Chicago. Fortunately, they were very flexible and I have worked in a telecommuting capacity ever since joining the magazine." The arrangement—while not so uncommon today—was considered a pioneering experiment back in 2003, and it has worked out well ever since.

STAYING INVOLVED IN THE PROFESSION

For many years early in his career, he was very active in AIChE, serving on various national committees and as Chair of the New York Section. "When I was elected a Fellow of AIChE—a pretty substantial honor—I believe at the time I was the youngest Fellow ever elected."

"When I was growing up, finances were an issue, so being accepted to The Cooper Union (which, at the time, was tuition-free), was a lifeline in terms of being able to go to college," he explains. "The Cooper Union specializes in art, architecture, and engineering, and I'd always been good at math and science, so I gravitated toward engineering." He initially opted for electrical engineering, "but then switched my focus to chemical engineering, which was more to my liking," he notes.

"I didn't realize at the time that Chemical Engineering graduates tended to have the highest starting salary among the various engineering disciplines," says Rosenzweig. "Rather, I decided to switch majors because dealing with chemicals and the equipment to make them seemed much more tangible and interesting than working on electronic circuits. I certainly don't regret my decision."

"The one thing I remember people always saying about chemical engineering—and their observations still hold today—is that the courses you can take in pursuit of your degree are very broad and quite flexible, and they give you a strong foundation to explore many different types of jobs and many different industry segments," he notes.

"When I graduated, and indeed for most of my career, the traditional chemical industry was full of chemical engineers, but today, many chemical companies have reduced the number of chemical engineers they have," he observes. Fortunately, the number of non-traditional scientific areas and industries that value—and indeed depend on—chemical engineering input has continued to grow. "Today, exciting career options exist for chemical engineers, in a growing number of non-traditional fields and industries, including nanotechnology (which has implications for pharmaceutical manufacturing, materials science, catalysis, and more), pharmaceuticals development, alternative energy, biotechnology, and more.

"Similarly, we're seeing tremendous investment in the pursuit and commercialization of new and non-traditional feedstocks," he says. "Chemical engineers are well-positioned to lead cross-cultural and interdisciplinary breakthroughs."

ACADEMIA—OFTEN LACKING IN TRUE INDUSTRY PERSPECTIVE

Speaking of today's undergraduate and graduate school experience, Rosenzweig worries about the "silos" that exist within many chemical engineering classrooms. "When I graduated, many of my faculty had good, practical industry experience, which helped to inform their classroom lectures in really meaningful ways. But that has changed dramatically in recent years," he laments. "Today, many faculty have been in academia their whole career and are not able to bring in any real-world perspective from having worked in industry."

"My advice for students: pay particular attention and give value to what the so-called 'adjunct professors' do. These instructors generally have spent most of their time in industry, and teach courses based on their particular area of expertise," explains Rosenzweig. "Many of them are, first and foremost, working chemical engineers—and, secondarily, instructors for a very specific class or course. Such practitioners often bring the type of real-world, first-person perspective from their industrial experience that 'more esteemed' lifelong tenured faculty can't provide."

As for his own career spent developing and improving editorial materials aimed at helping chemical engineers to do their jobs better and understand the new technologies and business trends that are shaping the industries in which they work, Rosenzweig has been very satisfied. "The field of technical editing is not an obvious career option for graduating chemical engineers and, quite honestly, the number of positions for engineers is shrinking. However, if someone goes in 'with eyes wide open,' the career path has a lot of advantages and can be quite intellectually satisfying," he notes.

"Being an editor, I've never really been pigeonholed into being a technical specialist in a very narrow area," explains Rosenzweig. "Rather, I'm always handling a variety of topics related to both established and new topics, in an array of areas, so the work never gets dull." ■

REFERENCES

"2017 AIChE Salary Survey," *Chemical Engineering Progress*, pp. 14–23, June 2017.

American Chemistry Council, www.americanchemistry.com/Jobs/ (accessed June 15, 2017).

Joshi, Amanda, "Salary and Job Satisfaction," *Chemical Processing*, pp. 18–24, April, 2017.

Statista, The Statistics Portal, "2016 Ranking of the Global Top 10 Biotech and Pharmaceutical Companies Based on Employee Number (in thousands)," www.statista.com/statistics/448573/top-global-biotech-and-pharmaceutical-companies-employee-number/ (accessed June 7, 2017).

Swift, Thomas Kevin, "2017 Guide to the Business of Chemistry" and 2018 private communication, American Chemistry Council, http://goo.gl/QBGtXT (accessed June 7, 2017).

US Bureau of Labor Statistics (BLS), Federal statistics on chemical engineers. www.bls.gov/ooh/architecture-and-engineering/chemical-engineers.htm (accessed on June 1, 2017).

Basic Concepts: Chemical Processes and Unit Operations

4

CHEMICAL PROCESSES AND UNIT OPERATIONS

The terms *chemical process*, *chemical processing*, and *chemical process industries* (CPI) are frequently used throughout this book. So, what is a chemical process? (see Figure 4.1)

A *chemical process* is a combination of steps in which starting materials are converted into desired products using equipment and conditions that facilitate that conversion (Solen and Harb 2012).

A *unit operation* is a basic step in a process. Unit operations involve either a physical change, such as evaporation, filtration or crystallization, or a chemical transformation, such as polymerization, isomerization or oxidation.

Many of the products used by modern society involve chemical or biomolecular processing in their creation. The processing steps or unit operations usually lead to physical and/or chemical changes. Below are a few examples.

Example 1. Manufacture of table salt (sodium chloride) from sea salt

This is an example of a chemical process that has been in use for thousands of years. It illustrates the use of a series of chemical engineering processing steps, or *unit operations*. Seawater (which contains 3.5% sodium chloride, plus smaller amounts of other salts such as calcium sulfate and magnesium chloride) is evaporated (unit operation: *evaporation*) to produce brine by solar heating (unit operation: *heat transfer*) in shallow ponds in an arid environment (see Figure 4.2). Sodium chloride crystals form (unit operation: *crystallization*), settle out (unit operation: *sedimentation*), and are separated from the brine and dried (unit operation: *drying*). In this example, mass transfer occurs as a result of the separation of liquid (water) in the seawater to the atmosphere as water vapor, leaving behind a salty brine.

Heat transfer occurs as solar radiation acts on the seawater/brine ponds. Crystallization of sodium chloride occurs when enough water has been evaporated from the seawater to form a concentrated brine in which the sodium chloride concentration exceeds its solubility limit. The salt crystals that form have a density that is about twice that of the surrounding briny (salty) liquid, so they settle to the bottom of the pond by gravity (by way of sedimentation). The salt crystals that form are separated from the brine. The other salts (such as calcium and magnesium salts) remain dissolved in the brine.

Example 2. Neutralization of an acidic waste stream

An aqueous acid waste stream contains hydrochloric acid, which must be neutralized with aqueous sodium hydroxide as a part of the treatment of the waste.

(4.1)   acidic waste (HCl) + caustic (NaOH) > salt (NaCl) + H_2O

The acidic waste is produced continuously, so a constant stream of aqueous sodium hydroxide must be added to the acidic waste in a tank. The tank is stirred by a mechanical agitator (unit operation: *mixing*) to bring the acid and the caustic into intimate contact.

FIGURE 4.1   In a typical chemical process plant, such as the one shown here, chemical engineers get involved in countless activities, ranging from selecting and managing raw materials and other key feed streams to designing, operating, and maintaining the chemical process operations that are used to convert starting materials into finished products. They are also intimately involved in the pollution-control systems and other environmental and safety systems that are required to ensure safe operation, and in the automation and control systems that are at the heart of all modern industrial facilities. When biological processes are involved in the chemical process operations or in the treatment of wastewater and other waste streams, biomolecular engineering graduates will also find many diverse opportunities for employment. (Source: Shutterstock ID 158232665)

FIGURE 4.2   Salterns (also called salt-evaporation ponds) are still used to promote the evaporation of water from seawater, allowing operators to harvest the sodium chloride and other salts that were dissolved in the seawater in solid (crystalline) form. (Source: Shutterstock ID 609481760)

The hydrochloric acid reacts with the sodium hydroxide via Equation 4.1 (unit operation: *reaction* or *neutralization*).

Note that in this scenario, the flowrate and concentration of the acidic waste stream varies over time, so it is essential that the flowrate of sodium hydroxide be adjusted to match the incoming waste (to ensure that the neutralization reaction occurs consistently). Although an operator could stand by the tank and continuously maintain the flow of caustic at the required rate, this would be inefficient and unnecessarily tedious.

Instead, the caustic addition is controlled automatically. This is done by continuously measuring the concentration of the unneutralized acid in the mixed tank using a pH meter. A flow-control valve then adjusts the flowrate of sodium hydroxide solution to maintain neutral pH (using *process control practices*) in the stirred tank.

Example 3. Combustion of natural gas: A gas-to-gas chemical reaction process to produce energy in the form of heat

The main component of natural gas is methane (CH_4), the simplest *hydrocarbon* (compounds of *hydrogen* and *carbon*). The reaction of methane with the oxygen in air follows the following equation:

$$(4.2) \qquad CH_4(g) + 2\,O_2(g) \rightarrow CO_2(g) + 2\,H_2O(g)$$

This process releases considerable heat in this highly exothermic (heat-releasing) gas-to-gas reaction (see Figure 4.3). Other hydrocarbons in natural gas also burn by similar equations with significant heat release. Natural gas is abundant, and is an excellent fuel, although it is a fossil energy resource that produces greenhouse gases such as carbon dioxide and minor amounts of nitrogen oxide as unwanted byproducts.

A *greenhouse gas* is a gas that absorbs infrared radiation from the sun and from the surface of the earth, trapping the heat in the atmosphere and increasing its temperature, causing global warming.

Because methane is also a greenhouse gas—one that traps about 25 times more thermal radiation than carbon dioxide—it is essential to eliminate natural gas leaks during the production and use of natural gas.

The heat that is produced during the combustion of natural gas is often used to evaporate water (a physical unit operation) under pressure to produce high-pressure steam. This high-pressure steam can then be used for a variety of industrial purposes, including—importantly—the operation of turbines that drive generators to produce electricity.

FIGURE 4.3   When natural gas (whose largest component is methane) is combusted with oxygen in gas-fired burners (such as the one shown here, inside a furnace firebox), energy is produced in the form of heat. (Source: Shutterstock ID 908262286)

Example 4. Combustion of gasoline: A gas-to-liquid chemical reaction that releases heat and produces mechanical energy

Gasoline is a mixture of liquid hydrocarbons, with octane (C_8H_{18}) as one of its primary components. The combustion of gasoline in air in a car engine follows the following equation:

$$(4.3) \qquad 2C_8H_{18}(g) + 25\,O_2(g) \rightarrow 16\,CO_2(g) + 18\,H_2O(g)$$

INDUSTRY VOICES

"Chemical engineering graduates should—if possible—try to start out at a plant-level job, in an operating company in the oil-and-gas, petrochemical, chemical or pharmaceutical industries, to get a real understanding of the day-to-day operations and maintenance. There is so much to learn from operators, maintenance technicians, engineers and supervisors who work in the plant."

—**Kathy Shell,
Applied Engineering
Solutions, Inc.,
page 56**

A large amount of heat is released by the combustion of the hydrocarbons in gasoline. This heat causes the combustion gases that are produced via Equation 4.3 to expand to a much greater volume than the incoming gasoline and air. The expansion of these gases is used to push the pistons in the cylinders of the engine, powering a car. The heat of the reaction (combustion) is also used to evaporate the incoming liquid gasoline fuel in the engine. In addition to nitrogen and carbon dioxide, the engine exhaust that is produced includes trace amounts of unburned hydrocarbons and nitrogen oxides, which are regulated air pollutants. Catalytic converters (designed by chemical engineers) are used to convert these potential pollutants (in the automobile exhaust) into water and carbon dioxide.

Example 5. Combustion of coal in air: A gas-to-solid chemical reaction that produces heat and electricity

Coal is a solid mixture of inorganic materials (which form ash during combustion) and complex hydrocarbons with roughly equal numbers of carbon and hydrogen atoms. (A typical empirical formula for the hydrocarbon fraction of bituminous coal is $C_{137}H_{97}O_9NS$.)

Solid chunks of coal are first ground (unit operation: *grinding* or *size reduction*) into a fine powder. The coal powder is then mixed (unit operation: *mixing*) with air at high temperature (unit operation: *heat exchange*) to give the simplified reaction of coal with oxygen in air (unit operation: *reaction* or *oxidation*), through the following equation:

$$(4.4) \qquad 2n\,CH(s) + 2.5n\,O_2(g) \rightarrow 2n\,CO_2(g) + n\,H_2O(g) + ash$$

where n = approximately 100.

The reaction that proceeds via Equation 4.4 releases considerable heat, as it is a highly exothermic gas-to-solid reaction. The combustion of coal also releases gaseous nitrogen oxides and sulfur oxides, and leaves behind a large amount of solid ash that may contain arsenic, mercury, lead, and heavy metals, many of which are toxic substances. Because coal has a higher carbon-to-hydrogen ratio than natural gas, combustion of coal burns more carbon and less hydrogen, and thus releases more carbon dioxide than natural gas for the same amount of energy produced.

CONSERVATION OF MASS AND ENERGY

One of the fundamental tenets of chemical and biomolecular engineering is that mass is conserved and energy is conserved. This means that chemical reactions are *balanced*, with the number of atoms of each element in the reaction mixture unchanged by the reaction. Thus, in Equation 4.3 above, for example, there are 16 atoms of carbon in both the reactants and the products. Similarly, there are 36 hydrogen atoms and 50 oxygen atoms on both sides of the equation. Because the number of each type of atom is the same on both sides of Equation 4.3, it is said to be a *balanced reaction*.

Mass is neither created nor destroyed. This statement is only true as an approximation, because the amount of mass change in *atomic* chemical reactions that involve changes in the chemical bonds in combinations of atoms is too small to measure. Albert Einstein's famous equation (Equation 4.5), applies rigorously to the transfer of mass to energy:

(4.5) $$E = mc^2$$

where
  E = energy
  m = mass
  c = speed of light

By contrast, the conversion of a small amount of mass through *nuclear* fusion or *nuclear* fission produces an enormous amount of energy. For example, the fusion of two hydrogen nuclei into a helium nucleus is the nuclear reaction by enormous amounts of hydrogen that powers the sun. While nuclear engineering is a viable career path for chemical engineers, nuclear fission and fusion reactions do not occur in a rate high enough to cause significant amounts of mass loss or gain in the chemical process industries.

Instead, for practical purposes, mass and energy are conserved in chemical and biomolecular reactions. By way of explanation: in most chemical reactions, there is either a release of heat (exothermic reaction) or an uptake of heat (endothermic reaction). In an *exothermic reaction*, heat is released as the reaction proceeds from reactants to products. By contrast, in an *endothermic reaction*, energy must be provided from the environment to drive the reaction forward. The magnitude of the heat of a reaction is the same for the forward and the reverse reactions; only the direction of energy transfer to or from the surroundings differs.

The concept of the conservation of mass and energy in chemical and biomolecular processes provides the chemical engineer with a valuable framework to use in the analysis and design of chemical processes. Equations 4.6 and 4.7 (below) quantify the conservation of mass and energy.

For continuous flow to and from a given vessel or container, the overall mass or material balance equation is:

(4.6)   total mass input = total mass output + total mass accumulation

The energy balance is shown using Equation 4.7:

(4.7)   total energy content of input mass to container = total energy content of output mass
       from container + heat released or consumed by reaction + accumulation of energy in
       container + energy transfer to thecontainer

There are similar equations for conservation of the mass of each of the individual elements in the reactants and products. Equation 4.8 is an example for the element carbon in Equation 4.3:

(4.8)   mass of carbon in octane consumed = mass of carbon in carbon dioxide created

This family of equations enables an engineer to predict the quantitative outcome of a chemical or biochemical process based on knowledge of the properties of the chemical reactants and products, the heat of the reaction, and the properties of the container.

This ability to predict outcomes based on a fixed set of principles enables the engineer to choose the best conditions of temperature, pressure, composition, and flowrates of the inputs to the process equipment, and the reaction conditions to produce the output stream at the desired temperature, pressure, and state of matter (for example, gas, liquid, or solid).

FLUID FLOW

INDUSTRY VOICES

"My role is to help the team address some nagging solids-processing challenges. You don't want to end up with a product that you cannot make, or a product you cannot handle in safe, effective processes, or products that no customers will pay for."

—Shrikant Dhodapkar, The Dow Chemical Co., page 60

WHAT ARE FLUIDS?

"A *fluid* is a substance that does not permanently resist distortion. An attempt to change the shape of a mass of fluid results in layers of fluid sliding over one another until a new shape is attained. During the change in shape, shear stresses exist, the magnitudes of which depend upon the viscosity of the fluid and the rate of sliding, but when a final shape has been reached, all shear stresses will have disappeared. A fluid in equilibrium is free from shear stresses." (McCabe, Smith, and Harriott 2004.)

While many people think that fluids only exist in liquid form, the truth is that most fluids can exist as either liquids or gases. Additionally, *supercritical fluids* exist between liquid and gas states and occur at temperatures and pressures above the critical point of the fluid in question. The critical point is a fundamental property of any molecule, and is the point where the phase boundary between liquid and gas disappears.

The properties of fluids are important because many chemical processes occur in the liquid and/or gaseous state. Because fluids can easily flow through pipes and mix with one another, many chemical processes are operated at pressure and temperature conditions that were chosen to ensure that the process materials would be in a fluid (not solid) state.

The molecules of a gas are separate from one another and can easily move. At ambient conditions, gases have much lower densities than corresponding liquids. For example, at 25°C, liquid water has a density of 1,000 grams per liter. By contrast, water vapor at 25°C and atmospheric pressure has a density of only 2.07 grams per liter.

FLUIDS AND PRESSURE

The basic property of a static fluid is pressure. A static fluid exerts pressure on the container it is in due to the continuous impact of molecules in the fluid against the wall of the container. As the temperature of a gas in a closed container is increased, the molecules move faster and have higher kinetic energy when they encounter the wall or other molecules in the fluid, thus increasing pressure.

As mentioned above, gases have much lower densities than liquids. Thus, while the earth's atmosphere is miles high, the weight of the air above us only creates a pressure of 14.7 pounds per square inch at sea level. By contrast, the depth of the oceans creates pressures miles below the surface of the ocean that can reach thousands of pounds per square inch.

Gases can be compressed to smaller volume and higher density because of the space between gas molecules. If gases are compressed to a sufficient level (specific to the composition of the material), they transition to a liquid state (assuming the temperature is below the specific critical temperature that dictates the gas-liquid transition for a material).

The molecules of a liquid are in direct and continuous contact with other molecules in the liquid, so the motion of the molecules (and the liquid they comprise) is somewhat restricted. As a result, unlike gas, it is hard to compress liquid because of the resistance of the short-range repulsive forces of molecules.

The high density of liquids impacts their ability to move. Most liquids are more viscous than the gaseous state of the same material. Examples of liquids with relatively high viscosity—that is, those that flow sluggishly compared to, say, those that pour more freely—include glycerol, molasses, and honey. Examples of less-viscous liquids (those that can flow more freely) include water, gasoline, and isopropyl alcohol. By contrast, the low density of gases translates into lower viscosities and an easier ability to flow.

FLUID FLOW IN PIPES

It is often necessary during chemical processes to move fluids from one piece of equipment to another. An engineered piping array provides a conduit that connects the individual components of equipment and vessels. One engineering problem in chemical processes is determining the size of a pipe for a given flowrate of liquid or gas. Pipes that are too large are expensive; pipes that are too small may not allow the required flowrate and may consume more energy to overcome the high pressure drop.

Pressure drop is the decrease in pressure of a fluid as the fluid flows through a pipe, a channel, or any other conduit. When a fluid flows through a pipe, there is a resistance to flow that is caused by momentum transfer from the moving liquid to the stationery pipe wall. The resistance to flow results in a loss of pressure in the fluid.

The cost of piping and fittings is often one of the greatest expenses in the design and construction of a process plant.

PUMPS

Pumps, which are used to move liquids, are ubiquitous equipment components in chemical process operations. There are many different common pumps types, including centrifugal pumps (see Figure 4.4), positive-displacement pumps, and others. Liquids are said to be incompressible fluids because most liquids do not change volume significantly when the pressure of the liquid is changed. When a liquid enters a centrifugal pump, the pump motor provides energy to accelerate the liquid.

Centrifugal pumps have a vaned impeller that spins inside the pump casing. When liquid enters the pump inlet at the center of the casing, it is spun out to the edge of the casing by the spinning impeller. An opening on the outer edge of the casing acts as an outlet that allows the fluid to flow out of the pump and into a connected pipe (Figure 4.5) that connects the pump to the downstream vessel or equipment. Familiar examples of centrifugal pumps include those used in swimming pools and cooling water pumps used in automobile engines.

FIGURE 4.4 Pumps are essential process equipment components, which can be found throughout all types of chemical process plants. Pumps move liquids throughout the facility from unit operation to unit operation. Shown here is one type of pump (a centrifugal pump) and the motor that operates it at a water-treatment plant. (Source: Shutterstock ID 577068664)

FIGURE 4.5   This cutaway photo shows the inner workings of a single-stage centrifugal pump. Centrifugal pumps have a vaned impeller that spins inside a pump casing. The impeller imparts kinetic energy to spin the liquid through the pump and out of the exit to a connected pipe, which allows the fluid to flow to a downstream process or equipment component. (Source: Shutterstock ID 749934097)

Positive-displacement pumps have a cylinder that contains a piston. When the pump motor pulls the piston away from the end of the cylinder, the inlet valve opens, letting liquid flow in from the inlet piping. The piston then changes direction and pushes the liquid out through an outlet valve that opens to the outlet piping (Figure 4.6). Positive-displacement pumps are used when liquid must be moved to a much higher downstream pressure.

Pumps provide pressure energy that enables liquids to flow through pipes, equipment, and vessels. The selection of the type and size of a pump needed to convey a liquid stream in a chemical process is a routine engineering task. The engineer must calculate the pressure needed to overcome the frictional pressure loss in the piping (which impedes flow) and the pressure in the downstream vessel or equipment.

FIGURE 4.6   A liquid-plunger pump (shown here) is a type of positive-displacement pump. When used in remote, offshore oil-and-gas recovery operations, this type of pump is used to transfer liquid from the offshore platform to shore. (Source: Shutterstock ID 631035461)

COMPRESSORS

Compressors are used to move gases through piping, vessels, and equipment. Just as with pumps, two common types are centrifugal and positive-displacement compressors. Compressors operate on similar principles to pumps, but because gases are compressible fluids, compressors and pumps differ significantly in design detail (see Figures 4.7 and 4.8).

FIGURE 4.7   Shown here is a detailed cross-section of the impeller inside a centrifugal blower. Similar to the centrifugal pump, the centrifugal blower has an internal impeller that is spun by a motor. The spinning vaned impeller imparts kinetic energy to the gas flowing into the blower, accelerating the gas that then flows into a pipe or downstream equipment. (Source: Shutterstock ID 663192727)

FIGURE 4.8   Compressors are used to move compressible gases through piping, vessels, and equipment components. In some ways, they operate similarly to pumps (which are used to move liquids); however, because they handle gases, they often rely on different engineering designs. Shown here is a multi-stage, positive-displacement compressor that is used to move natural gas (methane) through a process plant. (Source: ID 109681469)

DISTILLATION

In his own words
"Applying engineering analysis
to solve challenging distillation problems"

HENRY KISTER, CHE, FLUOR CORP.

Most recent position:
Senior Fellow and Director of Fractionation Technology, Fluor Corp. (Aliso Viejo, CA)

Education:
- BE, Chemical Engineering and Fuel Technology, University of New South Wales (Australia)
- ME, Chemical Engineering, University of New South Wales (Australia)

A few career highlights:
- Chaired 16 distillation symposia in AIChE meetings between 1986 and 2017
- Elected Fellow of the AIChE (US) and IChemE (United Kingdom and Australia)
- Elected Member of the US National Academy of Engineering (NAE)
- Received the AIChE Founders Award for Outstanding Contributions to Chemical Engineering (2009)
- Serves on the Advisory Board of *Chemical Engineering* magazine
- Received the Personal Achievement Award in Chemical Engineering from *Chemical Engineering* magazine (2002)
- Received the AIChE Fuels & Petrochemicals Division Award for Contributions to the Advancement of the Fuels and Petrochemicals Industries (2010)
- Received AIChE's Clarence G. Gerhold Award for Contributions to Separation Technology (2003), and the George Lappin Award for meritorious service to AIChE technical programming (2012)
- Received the Engineers Australia/IChemE/RACI ExxonMobil Award of Excellence in Chemical Engineering (2014) for significant contributions to chemical engineering through innovations or publication
- Received the AIChE's South Texas Section Chemical Engineering Practice Achievement Award (2005)
- Received the IChemE's Gold Medal for the outstanding Chemical Engineering Achievement of the Year, Australia (1980), and the Hutchison Medal for Most Stimulating Paper (2003)

- Authored 112 technical publications, mainly in distillation/absorption technology, published in technical journals and/or presented at technical conferences
- Authored three books on distillation: *Distillation Operation* (Kister 1990), *Distillation Design* (Kister 1992), and *Distillation Troubleshooting* (Kister 2006)
- Authored the chapter on distillation and absorption equipment in Green's and Perry's 2018 "Perry's Chemical Engineers Handbook", the chapter on distillation in the *Kirk-Othmer Encyclopedia of Chemical Technology* (2018), and the chapters on trays and packings in Schweitzer's *Handbook of Separation Processes for Chemical Engineers* (Schweitzer 1997)
- Has presented his IChemE-sponsored continuing-education course "Practical Distillation Technology" more than 500 times
- Is a Chartered Engineer (ChE), the equivalent of a licensed Professional Engineer, in the UK and Australia

Henry Kister is Senior Fellow and Director of Fractionation Technology for Fluor Corp. For the past 30 years, he has developed broad and deep expertise—and a worldwide reputation as a leading authority—in all phases of distillation, including design, troubleshooting, start-up, operation, control, debottlenecking, and revamp of fractionation processes and equipment.

"At Fluor, I get involved in all phases of distillation. This includes state-of-the-art distillation process and equipment designs, both for new construction and revamps," says Kister. "In recent years, many of the projects I have worked on have involved revamps of existing distillation units. Typically, these units are already being pushed to the limit, but the owner wants to implement engineering and operational changes to push the systems even harder to get even more distillation capacity from the system. I also get extensively involved in troubleshooting and improving existing distillation systems designed by others, identifying their problem, debottlenecking them, and reinstating trouble-free operation."

Prior to joining Fluor in 1999, Kister was a staff consultant on fractionation for Brown & Root, and prior to that he worked for the Fractionation Research Institute (FRI), specializing in fractionation hydraulics. Kister's career began as a process engineer at chemical company ICI Australia, and he later moved to operations as a start-up superintendent.

APPLYING CONVENTIONAL TECHNOLOGY IN NON-CONVENTIONAL WAYS

Much of Kister's work consists of leading troubleshooting investigations to resolve mysteries, in order to improve or expand distillation tower performance. "Often a minor flaw creates a colossal problem," he says.

In a recent project, Kister helped a client overcome operating difficulties and increase production yield on a C_3 splitter distillation tower designed by others that is used to separate propylene from propane. "At the time, this tower—which is 309 feet tall and 28 feet in diameter—was the largest distillation tower built in the US," says Kister. "But from the very start, the trays operated inefficiently so the tower was not able to produce on-specification polymer-grade propylene. I was assigned to lead the investigation to figure out what was going on."

"The most important thing to do is to get out there and talk to the client, to find out what's really going on," says Kister. "You can't do meaningful troubleshooting work 'from the armchair.' You have to get out of your chair and get your hands on the equipment, do adequate diagnostic testing, take measurements to better understand and correctly diagnose the problem." He adds: "One good measurement is better than a thousand expert opinions."

In this case, Kister and his team used a non-conventional approach to get a better idea of what was going on inside the column. "*Qualitative* gamma scans are a common and invaluable diagnostic in distillation, much the same as X-rays are applied in medicine," he said, "but in this case, the qualitative gamma scans did not go far enough to identify the problem. We applied an advanced *quantitative* gamma scan technique that we developed. This breakthrough method identified the problem: that the state-of-the-art trays inside the distillation column experienced severe channeling because they were designed with too much open area."

"We worked together with the vendor and the client, put our heads together as a team, and 'blanked' the excess open area under a very challenging schedule and constraints. At the end of the day, we successfully modified the trays for this enormous tower and largely improved tray efficiency, allowing it to produce polymer-grade product at higher capacity than ever anticipated."

Some of the new troubleshooting techniques that he has developed, have gained wider use among practising chemical engineers. These include the use of quantitative gamma scans to identify and quantify maldistribution of flow in distillation column trays; (the resulting "Kistergrams" transform gamma scans into vapor and liquid flow profiles) or the start-up stability diagram that identifies downcomer unsealing.

DISTILLATION: A VERY UNLIKELY CAREER PATH

Reflecting on many formative experiences in his life, Kister says: "My life has been one big miracle, for which I am thankful to the Almighty. So many unanticipated events, and so many crossroads where I could have easily taken the wrong turn. One wrong turn and my career would have dead-ended. It was the divine intervention that had led me to where I am today, and I've never looked back."

Kister recalls that he "always liked chemistry and wanted to study something that would let him have a very practical career." He was first introduced to the concepts of distillation in the third year of his chemical engineering undergraduate experience. "My professor did not explain the concepts very well and I did not understand what was going on, so, naturally, I thought to myself, 'Well, distillation is one thing I'll never understand and will never really master.'"

He recoils, remembering a "Principles Exam" that had only four questions on it: "Two of the questions were on distillation. Even though we needed to answer three questions out of the four, I literally had to leave those two questions blank," he says.

After that third year of his undergraduate study, Kister—who was a so-called Shell Scholar because he was supported financially by Shell Co. during his undergraduate years—worked the summer at the Shell refinery in Sydney, Australia. Kister recalls with humor that he was assigned a distillation project to work on. "I got there and started talking to the other summer interns, saying, 'Does anybody want to swap projects with me?'" As it turns out, his supervisor would not allow him to switch, so he was stuck with the unwanted distillation project during that fateful summer.

Initially, it was a painful experience. "I had no idea what I was doing, so after a few weeks, I went back to my supervisor, and said 'I don't really understand what I'm doing here.' He suggested that I spend some time checking the simulations they had been carrying out (on the mainframe computer) against how the unit was actually performing, says Kister."

"Once I started comparing the actual numbers with the simulations, things started to come alive and the project became exciting," says Kister. "But I still had no idea what I was doing." At the end, he had to write a report and give a presentation—and breathed a sigh of relief when neither the folks at Shell nor their professor consultant tore it apart.

Earlier that year, Kister had some personal issues that caused his grades to plummet, and he was resigned to losing his Shell scholarship due to poor academic performance. To his amazement, he was notified that he would get to retain it. "On award night, the Shell representative told everyone that my summer work had solved a big problem for the refinery, calling it 'some of the best work done by a summer intern in the prior 10 years.' I was told that based on that merit, I would keep the scholarship," says Kister. "At that point, I realized that not only was I starting to like distillation, but it looked like distillation was also staring to like me," he adds, with a laugh.

"I was determined to go and learn more about distillation, so I pursued a Master's degree in distillation under the late Dr. Ian Doig, a top-notch academic advisor who had spent 20 years working in industry before coming to the university. He was a great mentor," says Kister.

Interestingly, his advisor, despite having deep expertise in industrial operations, was not necessarily a recognized expert in distillation. "I chose him as my supervisor because he was a very warm person and a great mentor," says Kister. "At that time in my life, I needed his dedication to help get me through. Everybody was saying 'You're crazy going with this guy for your graduate work, he's not a distillation expert,' but my Mom really noticed his warmth and she recommended that I work with him. She had a good sense of people and I respected her opinion, so I did it and it worked out perfectly."

MAKING AN IMPACT AROUND THE WORLD

Kister has been hailed as the world's foremost authority on distillation and absorption troubleshooting. "My father was a medical doctor, whose diagnosis ability was phenomenal, and who, at the age of 95, gained the record of the oldest practising doctor in the state of New South Wales in Australia. My mother was a psychiatrist who never failed to identify the correct root cause of a patient's problem, and who worked to her last day," he says.

"I am grateful that the Almighty blessed me with some of my parents' diagnostic abilities and their passion and dedication to their profession," says Kister. "It has also been very important to have had strong academic/engineering mentors who warmly encouraged my professional development along the way, and to be very active in building a community of technical peers (by joining engineering societies, participating in meetings, taking leadership roles, and so on)."

Early in his career with ICI Australia, he began in process engineering and was later promoted to start-up superintendent for distillation. "I asked my boss at the time if there were any good books on that topic and, without missing a beat, my boss said: 'If you find any, buy two copies—one for you and one for me!'"

"So I started talking to all the folks with the grey hair, writing notes to myself and asking lots of questions. I surveyed all the operators, engineers, supervisors, and everyone was willing to share their experiences," says Kister. "I started working on actual start-ups and I was learning a lot, developing my own experiences."

After two to three years in this role, he wrote a big paper that was only tentatively accepted to a prestigious conference. "After 35 pages, I thought I had not said enough, so I wrote another 100 pages. It became too big for the conference, so I redirected it to *Chemical Engineering* magazine, whom I expected to "*um* and *ah*" about publishing it. To my amazement, it was immediately and enthusiastically accepted. *Chemical Engineering* magazine published it as a seven-part series between May 1980 and April 1981. The series was very well received by the chemical engineering community, and it really created waves," says Kister.

One thing led to another, and based on that high-profile article series, Kister was invited to develop a two-day continuing-education distillation seminar in early 1980s (which later grew to three days). The accompanying manual that he developed for the seminar ultimately evolved into three 700-page books. "The other experts in the field tended to emphasize the fundamentals and theory, while I focused on the practical, troubleshooting, and 'how to' approach, which until then had been sparsely addressed. So my seminars and books filled a void and they really hit a nerve with the practising engineers."

Since introducing this pivotal "Practical Distillation Technology" seminar more than 33 years ago, Kister has presented it more than 500 times in 26 countries on all six continents—averaging about 15–20 times per year, mostly onsite at different operating companies throughout the chemical process industries. He also presents public sessions of this IChemE-sponsored seminar once per year in Houston and London, and once every few years in centers in Europe, Asia, and Australia, working through local engineering societies.

Kister stresses the importance of building one's engineering knowledge in many ways. "If you learn from your own mistakes, you are smart," he says. "If you learn from other people's mistakes, you are wise." ■

ENGINEERING MANAGEMENT AND PROCESS SAFETY CONSULTING

In her own words
"Emphasizing process safety throughout the entire organization"

KATHY SHELL, PE, APPLIED ENGINEERING SOLUTIONS, INC. (aeSOLUTIONS)

Current position:
Executive Vice President, Process Safety and Lifecycle Solutions, Applied Engineering Solutions, Inc. (aeSolutions; Greenville, SC)

Education:
• BS in Chemical Engineering, University of Akron (Ohio)

A few career highlights:
• Has served as Executive Vice President, Process Safety and Lifecycle Solutions, of Applied Engineering Solutions since 2017
• Has served as Chief Executive Officer of Applied Engineering Solutions since 2014
• Served as Senior Process Safety Director of Applied Engineering Solutions from 2010 to 2014
• Served as Business Manager (Ohio and Philadelphia Offices) and Process Safety Practice Leader for consultancy RMT, Inc. from 1997 to 2010
• Served as Process Engineering Manager, Thermal Engineering Manager, Detailed Design Manager, and Turnkey Incineration System BD Manager for International Technology Corporation, Inc. from 1985 to 1997
• Served as a production engineer for Dow Chemical Company from 1983 to 1985
• Is a licensed Professional Engineer (PE)

"I started out my career in a co-op program for Diamond Shamrock, while I was still in college," says Kathy Shell. "That was a valuable experience as it gave me exposure to chemical engineering careers in design, production, laboratory, and process development." Upon graduation, Shell took on a plant engineering position with Dow Chemical Corp. (Midland, MI). "I gained foundational experience there in process design, maintenance, operating discipline, loss prevention principals, and capital project management." For personal reasons, Kathy left that position and relocated to a former Dow engineering consultancy office owned by IT Corp. (now IT Shaw; Knoxville, TN), joining that company as a process engineer.

Shell was promoted to her first manager position (over the Process Engineering Department) in 1987, and she remained with IT Corp. for ten years, moving up through various group-management positions. This allowed a percentage of time for client-relationship development and technical project leadership.

She moved from that position into an engineering consultancy leadership role with RMT, Inc. (Columbus, OH) in 1997. Eventually, Shell assumed the role of Business Manager over the company's Ohio and Philadelphia offices, collaborating with staff consultants in the environmental, health, and safety arena. Concurrently, she continued to lead a national offering in process safety.

HER LATEST FOCUS

Eventually, in 2010, Shell joined Applied Engineering Solutions, Inc. (aeSolutions), a process safety, engineering, and automation company, as the Director of Process Safety. She assumed the role of CEO in 2014 and later transitioned to Executive Vice President of Process Safety and Lifecycle Solutions.

"Our vision is to improve the process safety performance of our clients," says Shell. "My efforts are primarily focused on several areas—strategically aligning the company resources to achieve that vision, building valued relationships with our clients, promoting a positive work environment for our employees, and delivering a return on investment to our shareholders."

"When I first joined the company, my goal was to expand the process safety consultancy and safety-instrumented system (SIS) engineering services that we provided to clients throughout the US, she adds."

Shell notes that the company brand was initially recognized locally in Greenville (for providing system integration services) and Anchorage, Alaska (for providing process safety and SIS services). "Our plan has been to increase industry recognition of aeSolutions as a provider of process safety lifecycle services across the entire US."

"It's been a professionally rewarding journey thus far," says Shell. "I have integrated my chemical engineering principles and early plant experience into a lifelong career in process engineering, project management, process-safety consulting, and business management. My career started with receiving training in loss-prevention principles from Dow Chemical, and expanded into assisting clients with achieving the risk-management objectives of OSHA's Process Safety Management and US EPA's Risk Management Program Standards, 29 CFR 1910.119 and 40 CFR 68."

THE HALLMARK OF ANY CHEMICAL ENGINEERING DEGREE: FLEXIBILITY

"A chemical engineering degree provides a great foundation for working with industries that use and manufacture hazardous chemicals, and allows you to be deeply involved in process safety management (PSM)," says Shell. She notes that the principles and practices of PSM are wide-ranging, calling for, among other things, the development and implementation of management systems, procedures, and tools that can help to ensure the safe handling of hazardous chemicals. "This involves the application of inherent safety principles, process hazard analysis, proper equipment design, establishing a hierarchy of controls, and putting the putting the practices in place to ensure their integrity," she says.

Shell suggests that chemical engineering graduates should—if possible—try to start out at a plant-level job, in an operating company in the oil-and-gas, petrochemical, chemical or pharmaceutical industries, to get a real understanding of day-to-day operations and maintenance.

"Start with 'boots on the ground' and then branch out or specialize from there," she says. "There is so much to learn from operators, maintenance technicians, engineers, and supervisors who work in the plant."

After developing her early experience in plant engineering, process engineering, and operations management, she began to focus on process safety, "as it is fundamental to all chemical process operations," she says. Shell notes that people don't always fully appreciate that embracing a rigorous approach to process safety involves much more than just engineering solutions to ensure the safe handling of materials. "I've always kept a foot in the technical realm, so I can maintain my aptitude and my appreciation for the effort required to execute work flow, but I've also focused on tying process safety concepts together with our clients' business objectives," says Shell.

"The goal is to implement a rigorous program, to design, operate, and maintain all systems in the safest possible manner, to install and operate the right monitoring devices and systems to inherently reduce the risk associated with process operations, and to have sufficient support from all levels of personnel throughout the organization."

"When companies do this, they are able to more predictably understand, control, and reduce risk, to manage change, and to optimize processes and operate the plant more efficiently," she adds.

Shell's company creates software that supports the full, sustainable implementation of the functional safety-lifecycle standards ANSI/ISA 84.00.01 (modified IEC 61511), helping to provide connectivity to the data for all stakeholders. Specifically, this includes:

- Identification of potential hazards
- Design of engineered safeguards
- Management of independent protection layers
- Development of proof-test plans
- Management of reliability data for the instrumented functions
- Reporting of health metrics
- Demand/cause tracking that is necessary to close the loop

"The important thing is to empower personnel with roles and responsibilities across the lifecycle, using reliable data in a timely fashion. This ensures that there's only 'one version of the truth,' and that all stakeholders can make informed decisions at all times," she says.

She notes that today's chemical engineering graduates are more savvy and comfortable with software database tools. Such models and tools are designed to support and optimize the lifecycle workflow, from risk analysis and engineering design, to operations and maintenance, reporting, controlling and auditing, and verification. "The next generation of engineers will be able to teach us even more, and be more savvy in the development and use of these advanced tools, and the ability to leverage data in the most advantageous way."

GETTING HIGH-LEVEL BUY-IN IS KEY

"In recent years, I've moved into providing process safety workshops for executive-level employees at industrial companies, who can have a far-reaching impact on how the company embraces process safety management (PSM)," says Shell. "PSM is so closely entwined with the notion of operational excellence, so empowering these individuals with a greater understanding of how their actions, or lack thereof, foster a culture of mindfulness and recognition of the business value of process safety management principles and practices is very rewarding."

The payoff from an investment in process safety comes with improved quality, productivity, profitability, and reputation, that is, the ability to combine operations and business excellence. These companies are more effective at managing high-hazard processes with low risk to their employees, the general public, and the environment.

RECALLING HER INTRODUCTION TO CHEMICAL ENGINEERING

Shell was the Valedictorian in her high school class. "I thought about being pre-med in college, but I really liked and respected my high school Chemistry teacher," says Shell. "I remember her explaining to all the boys in the class that if they really liked chemistry and math, they should pursue chemical engineering as a college major. As I sat there, listening to her counsel, I thought 'I can do that, too!'" she says. She looked into the interesting and varied course work associated with the chemical engineering major, the flexible career paths and strong compensation that chemical engineering graduates enjoy—and the rest, as they say, is history.

Of the 22-or-so students in her chemical engineering graduating class, only four or five were female. "I didn't really look at it as being unique at the time, but I started noticing that the women's bathrooms were never near the engineering offices," Shell recalls.

That said, she notes that "throughout my career, at every hurdle I faced, I always found people there to pull for me and doors that opened up," says Shell. "Engineers are problem-solvers by nature. They are always open to learning your circumstances and helping you to get through things and move to the next level. So don't be afraid to ask for help, seek counsel from peers and mentors, and partner effectively with colleagues."

While in college, Shell had a co-op job at Diamond Shamrock. "It gave me a chance to work in a developmental lab, in R&D, in a process plant, and in other assignments," says Shell. "It brought clarity and passion to my undergraduate experience and solidified my choice to stay put in chemical engineering (as opposed to a possible switch to architecture, which I had thought about for a while)."

"It's no surprise the rigorous chemical engineering curriculum requires focus and commitment," she says. "But a chemical engineering degree also provides a lifetime of opportunities to be intellectually challenged—it's creative work that's never boring, you can choose numerous career paths, you can work in a variety of industries, and you will be beneficially compensated," she says. "There are not many degrees that give you as much security, flexibility, and personal satisfaction." ■

PARTICLE TECHNOLOGY AND SOLIDS PROCESSING

In his own words
"Applying fundamentals of particle technology and innovative approaches to solve challenging industrial problems, from conception to scale-up"

SHRIKANT DHODAPKAR, PhD, THE DOW CHEMICAL COMPANY

Current position:
Fellow, Performance Plastics Research and Development for The Dow Chemical Company

Education:
- BTech, Chemical Engineering, Indian Institute of Technology (New Delhi, India)
- MS and PhD, Chemical Engineering, University of Pittsburgh (Pittsburgh, PA)

A few career highlights:
- Fellow of the American Institute of Chemical Engineers (AIChE)
- Former Chair of the AIChE's Particle Technology Forum and served on the AIChE's Chemical Technology Operating Council
- Former Technical Program Chair for the Fifth World Congress on Particle Technology
- Co-author of one or more chapters in five professional handbooks, including Perry's Chemical Engineers' Handbook, 9th Edition, (2017), and author of more than 40 published technical articles and 100 peer-reviewed internal Dow research reports
- Recipient of 15 Dow Technology Awards (which recognize individuals who have successfully implemented innovative technology with demonstrated value in excess of US$10 million) and Dow's Excellence in Science award (which recognizes sustained accomplishments and value impact through application of science)
- Holds six patents (as co-inventor)
- Serves on the Editorial Advisory Board of *Powder and Bulk Engineering* magazine, and as the Editor of the *Particle Technology Forum* newsletter

The world of particle technology and bulk-solids handling may be a subsector of the overall chemical engineering profession, but it represents a universe all its own, encompassing many familiar and challenging processes and unit operations. Falling under this umbrella are such engineering endeavors as pneumatic conveying, gas-solids separation, fluidization, powder mechanics and granular flow, silo and hopper design, particle formation and engineering, coating and drying technology, mixing, blending and segregation, the impact of particle size and particle-size distribution, solid-solid separation, feeding, metering and dosing, design of solids-processing systems, processes to ensure safe storage and handling of powders (and all the dust-explosion hazards associated with powders), and much more.

According to Shrikant Dhodapkar, a Fellow with Dow Chemical Company's Performance Plastics Process Research & Development (R&D) division, roughly half of all final products created by the chemical process industry (as well as the raw materials used, and intermediate or byproduct streams produced) exist in particulate or solid form at some point or another, and thus require the use of the solids-based unit operations noted above. "Various studies during the past four decades have shown that these types of solids-processing unit operations are twice as likely to fail during start-up as compared to other unit operations—yet these unit operations are often seen as the 'back end' of the process, and are not given due consideration during process design."

"The chemical engineering curriculum tends to be so full at the undergraduate level that most students do not learn enough about particle technology, or even gain basic familiarization with unit operations related to solids processing," says Dhodapkar. "For many, that level of specialization does not come until graduate school—and only then if students choose to focus on it."

As a result, "our undergraduate students tend to be extremely unprepared to tackle the types of problems they will encounter, and many young chemical engineers enter the workforce without sufficient understanding or training in these key aspects of chemical process operations," he says.

Dhodapkar has worked hard throughout his career to help address the "significant gap between how we are training our undergraduates and what many graduates will face when they go to work, with regard to handling bulk solids during process operations." He does this through a variety of means. For instance, he has always been very active in the AIChE and other professional organizations, has published technical articles and written book chapters—in collaboration with other experts in the field, with the express objective of disseminating knowledge and experience—and continues to teach courses and workshops within Dow and externally.

The ability to "share the wealth of knowledge" and help other students and technical professionals to improve their understanding of complex solids-handling challenges is one aspect of his role at Dow and his overall career that Dhodapkar particularly enjoys. Toward that end, he developed a series of comprehensive solids-handling courses (covering 12 major topics in particle technology), which he has presented to Dow engineers all over the world.

"I'm happy to see that while particle technology was often seen as peripheral to chemical engineering over the years, it has recently become much more integral, but there is still much work to be done to be sure all engineering students and young engineers have sufficient grounding in key aspects of bulk-solids processing, he says."

THE VALUE OF CREATIVITY AND INNOVATIVE PROBLEM-SOLVING

In his role as a Fellow with Dow's Performance Plastics Process R&D division, Dhodapkar functions as an "internal consultant" to Dow's production plants all over the world, helping with design, installation, and troubleshooting on a wide variety of issues that arise related to solids handling and processing. He is also responsible for discovery, innovation, and process development for technologies related to solids handling and processing. He finds particular satisfaction in solving difficult problems through creativity and innovative problem-solving approaches, and bringing them to practice.

Dhodapkar started his career with The Dow Chemical Company about 25 years ago in the Core R&D division. "I've always functioned as a researcher in applied engineering, and I've enjoyed every moment of it," he says.

His PhD was focused on particle technology with a skill set Dow was specifically looking for. "At Dow, I had the opportunity to collaborate with brilliant scientists and engineers, and its global footprint and vast resources provided a large sandbox to play in," he adds. "And it's been a great journey ever since, even though my role has continued to evolve over time."

"I develop technologies to address gaps that arise during our process operations, and help to create and serve as a repository for excellence and experience in a particular area, such as bulk-solids processing and particle technology," he says. "And it's worth noting that you don't always need to reinvent the wheel every time—there is a tremendous amount of knowledge out there, so often it's just a matter of being aware of what's going on throughout the industry (knowing the history and keeping up with ongoing advances) and then connecting the dots."

"I love being involved in all stages of a project—from conception to execution," says Dhodapkar. "If a chemist walks into the room and says 'I've got this exciting molecule and I can do all this exciting stuff with it, but I've only produced ten grams of it so far,' we then get to work to answer the questions, such as 'how do we scale it up, demonstrate it at the pilot-plant scale, produce it at a commercial scale, and get customers to buy it?'"

"My role is to help the team address some nagging solids-processing challenges," he adds. "You don't want to end up with a product that you cannot make, or a product you cannot handle in safe, effective processes, or products that no customers will pay for. If you're not making money then you're just having fun—but as a developer and manufacturer of chemical products, you need to do both."

Dhodapkar notes the importance of interdisciplinary work on thorny projects. "There's always a feedback loop—you learn a lot along the way with any project, and there are always certain things you wished you'd done or done differently, or questions you wish you'd asked, or ways in which you would have directed the project better," he says. "The goal is to capture these insights and use them to improve future endeavors."

CONSIDER YOUR OPTIONS

For young graduates figuring out their possible career paths, the question will arise as to whether to pursue a position at a larger company or a smaller company, and each has its pros and cons. "I always encourage young engineers to start their careers working in areas that have the broadest possible scope, and then they can move toward their specialization as they grow," says Dhodapkar.

"In general, larger companies typically provide opportunities to be exposed to a wide range of people, technologies, equipment, and development resources, including ongoing engineering courses and 'soft-side' training related to such things as public speaking, software training, and more," he notes. "By contrast, working in a smaller company, you may not have access to as many resources, but you are more likely to hold more roles and add responsibility more quickly, and to touch more aspects of the business."

"At the end of the day, engineering students should understand their strengths and proclivities and should follow those instincts," he says. "I like to see an inherent curiosity and a real passion in the new graduates we hire. They should exude unbridled enthusiasm, willingness to explore (and sometimes fail) various facets on their own and be willing to challenge the status-quo."

Similarly, he strongly suggests that young engineers seek and find mentors within their organization, and encourages seasoned engineers (and corporate management) to help connect younger engineers with more experienced ones. "Given the time and work pressures and productivity challenges that many working engineers face today, classical mentoring seems to have fallen off a bit, but it's still important to find ways to make it happen as it benefits all parties, helps to nurture the newer engineers, and helps to ensure that institutional knowledge is passed down."

Dhodapkar always offers the following two pearls of wisdom to his young engineering colleagues.

- *No effort ever goes wasted, so don't get discouraged.* "The effort may pay off now, or it may pay off later—you never know. But if you don't put the effort in, you'll never have opportunities arise for you; too often, people get discouraged because they don't see instant rewards."
- *Opportunity always favors the prepared mind.* "Continue to push yourself to learn and grow and challenge yourself throughout your career."

ON CHOOSING CHEMICAL ENGINEERING

Dhodapkar grew up in India, and recalls that "in those days, you had to track your career into engineering or medicine early on, or you would not get accepted into the prestigious institutions." Both of his parents and his older brother are physicians, but, he says: "I was always that incorrigibly inquisitive kid who turned every appliance into an experimental project and loved creating 'Rube Goldberg'-type contraptions. I knew I wanted to be an engineer, not a physician."

At the Indian Institute of Technology in New Delhi, he enrolled into the chemical engineering program because it offered many more possibilities for future career paths. Chemical engineering is an amalgamation of numerous fields of science and engineering. "One of my professors there specialized in particle technology and did consulting work with the industry, and he always had interesting case studies to discuss," says Dhodapkar. "As I envisioned my future career, I wanted to find a space where I could make a difference. While the physics of fluids is reasonably well established, even at the molecular level, the behavior of particulates on micromechanical and bulk level is still an evolving science," he says, adding: "The technical opportunities in particle technology and bulk-solids handling, and their relevance to chemical, plastics, pharmaceutical, and food industries, provided the perfect opening that I was looking for."

He came to the U.S. for his Master's degree and PhD, because there was so much pioneering work being done here in the field of bulk solids and particle technology, and he knew there would be good companies looking for such experts. "During my graduate studies, I was very fortunate to have a thesis advisor who encouraged me to freely explore other topics and look for connections. While on these exploratory trails, I even took up experimental studies in electrochemistry and coal dehydrogenation, but I quickly realized that my true passion was particle technology—and has remained so, even after nearby 30 years." he says.

"During my graduate studies, I got acquainted with real-world industrial problems by collaborating with various external consultants in the field. Such collaborations made me appreciate the difficulties in applying theoretical concepts to solve real-world problems where uncertainty and incompleteness of input information must be dealt with. So, I learned and applied concepts from Fuzzy Set Theory and Expert Systems to bulk-solids handling Such cross-fertilization of ideas from different fields, and working at the interface of various technologies, is where I have always believed one could really make a difference," says Dhodapkar. "This has been a dream job for me, and there have been endless opportunities to contribute and grow within the company. I've really been lucky in that way."

Dhodapkar notes that the entire realm of chemical engineering continues to grow. "It's such a dynamic field and continues to gobble up adjacent, new, and allied fields—such as biological processes, nanotechnology, pristine processing (that is, the need to engineer ultra-clean processes and systems that are required for pharmaceutical, electronics, and food-and-beverage production) safety-related issues, environmental management, and more—so the umbrella just gets bigger and bigger, providing new and interesting avenues for chemical engineering graduates to explore," he says. "In this way, it's very different from other majors and from many of the other engineering disciplines." ∎

REFERENCES

Kister, H.Z. 1990. *Distillation Operation*. New York: McGraw-Hill, Inc.

Kister, H.Z. 1992. *Distillation Design*. New York: McGraw-Hill, Inc.

Kister, H.Z. 2006. *Distillation Troubleshooting*. Hoboken, NJ: Wiley-Interscience.

Kister, H.Z. Gas absorption and other gas-liquid operations and equipment, Section 14 in Don W. Green and Robert H. Perry (eds.) 2018. 9th Edition. Perry's *Chemical Engineers' Handbook*. New York: McGraw-Hill, Inc.

McCabe, Warren L., Smith, Julian C. and Peter Harriott. 2004. *Unit Operations of Chemical Engineering*, 7th Edition. New York: McGraw-Hill, Inc.

Solen, Kenneth A. and John N. Harb. 2012. *Introduction to Chemical Engineering*, 5th Edition. Hoboken, NJ: John Wiley & Sons, Inc.

Basic Concepts: Equilibrium and Rate Processes

5

THERMODYNAMICS

FUNDAMENTAL DEFINITIONS IN THERMODYNAMICS

Thermodynamics is a branch of physics that is concerned with temperature and heat, and their relation to energy and work in macroscopic systems (Van Ness and Abbott 2008). There are four laws of thermodynamics (discussed below) that apply to thermodynamic systems, such as heat engines or chemically reacting substances. There are numerous other equations that derive from those laws, and those equations allow users to predict the outcome of physical and chemical phenomena involving thermodynamic systems.

Chemical thermodynamics is the study of the interrelation of energy with chemical reactions or with a physical change of state, within the confines of the laws of thermodynamics. Numerous chemical engineering unit operations and processes can be represented as thermodynamic systems for the purposes of analysis and design.

Conceptually, in thermodynamics, a *system* may be an object, a quantity of matter, or a region of space that is selected for study and set apart (conceptually) from everything else; the "everything else" is referred to as the *surroundings* (Van Ness and Abbott 2008). The *boundary* of the system can be imagined as an envelope that encloses the system and separates it from its surroundings.

A system can interact with its surroundings in several ways. If there is no interaction, the system is said to be *isolated*. A system that is not isolated may allow the exchange of energy and/or material across its boundary. A system that allows exchange of matter is said to be *open*. If energy can exchange between the system and the surroundings, but no matter can be exchanged, the system is said to be *closed*.

Initially, changes in temperature or pressure may occur within an isolated system, even if there is no exchange of matter or energy with the surroundings. However, these changes eventually stop, and the system is said to be in a final static state of *internal equilibrium* (Figure 5.1).

For a closed system that is no longer exchanging energy with its surroundings, the final static system is said to be in *external equilibrium* (Van Ness and Abbott 2008).

Equilibrium is an important concept in thermodynamics. For a system that has internal equilibrium, its *identifiable and reproducible state* occurs when all of its properties, such as temperature, pressure, and molar volume, are fixed and cannot be altered without the introduction of some additional change.

When a system that is initially at equilibrium is displaced—for example, by a transfer of energy—it undergoes a process called a *change of state*. Once a new equilibrium is established, the identifiable and reproducible properties of the new state are achieved. If the process is conducted so that the state of the system is only changed by incremental amounts (and thus essentially remains at equilibrium throughout the process), the process is called a *reversible process* (Van Ness and Abbott 2008). In a reversible process, the direction of the change of state can always be reversed by a change of direction of the motive variables.

FIGURE 5.1   Internal equilibrium is like a scale that is balanced. (Source: Shutterstock ID 152090606)

THE FOUR LAWS OF THERMODYNAMICS

- *Zeroth law of thermodynamics:* If two systems are each in thermal equilibrium with a third, they are also in thermal equilibrium with each other (Moran and Shapiro 2008). The Zeroth law of thermodynamics leads to the conclusion that temperature is a measure of thermal equilibrium, and that if two systems are at the same temperature, then the small, random exchanges between them, such as Brownian motion, will not lead to a net change of energy (Moran and Shapiro 2008). This law also provides the basis for the design of thermometers.*

Consistent with the Zeroth law is a *first postulate of thermodynamics* (Van Ness and Abbott 2008):

There exists a form of energy, known as *internal energy* (**U**), which, for systems at internal equilibrium, is an intrinsic property of the system, functionally related to the measurable coordinates that characterize the system (Van Ness and Abbott 2008).

- *First law of thermodynamics:* The total energy of any system and its surroundings is conserved (Van Ness and Abbott 2008). The First law of thermodynamics is an expression of the principle of conservation of energy. As discussed earlier regarding Equation 4.7, energy can be transformed (changed from one form to another), but cannot be created or destroyed.

    When applied to closed (constant-mass) systems in which only internal energy (U) changes occur, the First law is expressed mathematically as (Van Ness and Abbott 2008):

(5.1) $$dU = dQ + dW$$

where:
   dU = the total differential change in internal energy of system
   dQ = the differential quantity of heat transferred to system
   dW = the differential quantity of work done on the system by the surroundings.

---

* The Zeroth law was discovered after the First, Second, and Third laws of thermodynamics had already been identified and established. Therefore, to avoid confusion and renumbering of the other three laws, it was numbered the Zeroth law (Moran and Shapiro 2008).

Note that internal energy is a state of the system, whereas heat and work modify the state of the system. A change of internal energy of a system may be achieved by any combination of heat and work added or removed from the system. Because internal energy is a state of the system, how a system achieves its internal energy is path independent.

The First law of thermodynamics is also the *second postulate of thermodynamics* (Van Ness and Abbott 2008). The *third postulate of thermodynamics* is as follows:

There exists a property called *entropy* (S), which, for systems at internal equilibrium is an intrinsic property of the system, functionally related to the measurable coordinates that characterize the system (Van Ness and Abbott 2008).

For reversible processes, changes in this property may be calculated by the following equation:

(5.2)
$$dS = \frac{dQ_{rev}}{T}$$

where:

dS = the change in entropy of the system

dQ_{rev} = the reversible transfer of heat into the system

T = the absolute temperature

- *Second law of thermodynamics:* Heat cannot spontaneously flow from a colder location to a hotter location (Van Ness and Abbott 2008). The Second law of thermodynamics is also the *fourth postulate of thermodynamics*, which is as follows:

The entropy change of any system, and its surroundings, *considered together*, resulting from any real process is positive, approaching zero when the process approaches reversibility (Van Ness and Abbott 2008).

This explains why, in nature, things tend to decay over time. Entropy is a measure of how much decay has occurred (Moran and Shapiro 2008).

- *Third law of thermodynamics:* As a system approaches absolute zero, all processes cease, and the entropy of the system approaches a minimum value (Moran and Shapiro 2008).

Absolute zero, at which all activity stops, is −273.15°C (Celsius), which is equivalent to −459.67°F (Fahrenheit) or 0°K (Kelvin). The Third law provides an absolute reference point for the determination of entropy. Entropy relative to absolute zero is absolute entropy (Moran and Shapiro 2008). The *fifth postulate of thermodynamics* is as follows:

The macroscopic properties of homogeneous PVT systems at internal equilibrium can be expressed as functions of temperature, pressure, and composition only (Van Ness and Abbott 2008).

Many fluids in chemical processes can be considered homogeneous PVT systems, and this enables the calculation of the properties of those fluids at a given state. This, in turn, enables prediction of the properties for purposes of design of these systems.

TRANSPORT OF MASS, HEAT, AND MOMENTUM

Transport of mass, heat, and momentum are important phenomena that take place in most unit operations, as well as in nature. While the principles of thermodynamics are used to describe equilibrium for systems, transport phenomena are non-equilibrium processes, or rate processes. The following are examples of transport phenomena, which are the direct result of molecular motion:

- Mass transfer, examples:
  - Dispersion of cream in tea
  - Condensation of water from warm, humid air onto cold glass

- Heat transfer, examples:
  - Heat from a stove onto a pot of water
  - Solar radiation on a sunny day
- Momentum transfer, examples:
  - Stirring liquid in a pot
  - Standing outside on a windy day

MASS TRANSFER

Mass transfer involves the transfer of molecules from one place to another. The simplest and the slowest mass transfer is by **molecular diffusion**. A drop of cream in an unstirred cup of tea diffuses into the tea, and the tea diffuses into the cream. To speed this process up, we stir the cup with a spoon—this is mass transfer by **convection**. Convection is the application of force (in this case, by the spoon we are stirring in the cup) to move clusters of molecules. Mass transport is important in the mixing of reactants in a reactor to ensure intimate contact between the reacting molecules. Mass transport is also very important in many widely used unit operations, such as distillation, absorption, and ion exchange. These unit operations will be described later in this book.

HEAT TRANSFER

Heat transfer can occur by **conduction**, **convection**, and/or **radiation**. In solids or stagnant fluids, heat is transferred by the comparatively slow process of conduction. With conduction, heat transfer occurs as more-energetic molecules (at higher temperatures) contact nearby less-energetic molecules (at lower temperatures), transferring kinetic energy to the slower and cooler molecules, and thus raising their temperature. As with mass transfer, heat transfer by convection is much faster because it usually involves flow and/or mixing processes that rapidly mix packets of fluid instead of just heat transfer at the molecular level between individual molecules.

Convection can occur as either *forced convection* or *natural convection*. In forced convection, a machine component is used to impart the mechanical energy needed to promote heat transfer. Examples include the use of a pump to move fluid through a heat exchanger or a pipe, or the use of a mixer in a tank.

By contrast, *natural convection* occurs because warmer fluids are usually less dense than colder fluids of the same composition. Thus, when fluids of different temperatures (and thus different densities) are co-mingled, gravity causes the colder liquids to sink and the hotter fluids to rise, providing mixing and heat transfer that are driven only by temperature and density differences.

Meanwhile, familiar forms of *radiant heat transfer* include solar heating and radiant heating that results from an exposed flame.

MOMENTUM TRANSFER

Momentum transfer can be illustrated by a fluid flowing in the x-direction parallel to a flat surface. Momentum transfer occurs in the fluid because the fluid at the plate surface has a velocity of zero, while the velocity of the fluid in the x-direction increases as the distance in the z-direction perpendicular to the plate increases. Each layer of fluid parallel to the plate has a momentum that is proportional to the velocity and the mass of the fluid element under consideration.

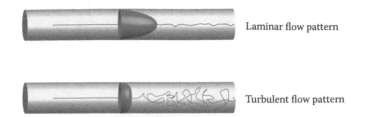

Laminar flow pattern

Turbulent flow pattern

FIGURE 5.2   Illustration of laminar and turbulent flow patterns in a pipe. At slower flow rates, fluid flow is laminar with fluid elements flowing parallel to one another with little radial mixing, and the velocity profile is parabolic. The fluid velocity in the center of the pipe is twice the average fluid velocity in the pipe. At higher flow rates, in turbulent flow, there is vigorous mixing across the radius of the pipe and the velocities across the pipe are much closer than in laminar flow. (Source: Shutterstock ID 510364918)

The fluid may be in either *laminar flow* or *turbulent flow*. In laminar flow, each layer of fluid parallel to the plate flows at a constant velocity. The velocity increases with increasing distance of the layer from the plate. Molecular diffusion of a faster layer of fluid into an adjacent slower layer of molecules will transfer kinetic energy to the slower layer. Similarly, molecules in the slower layer will diffuse into the faster layer, absorbing kinetic energy.

The result of this type of flow is the transfer from momentum in the z-direction to successive layers of fluid. At the plate/liquid interface, the momentum transfer from the fluid to the plate will cause a force on the plate in the direction of the fluid flow. In laminar flow of Newtonian fluids, the momentum flux in the z-direction will be directly proportional to the velocity gradient in the z-direction.*

In turbulent flow (in a vessel with the same geometry), the momentum transfer is so vigorous that eddies form to carry packets of fluid back and forth in the z-direction. This transfers momentum and energy much more rapidly than in laminar flow. In turbulent flow with the same system, the force on the plate from the fluid will be larger than in the laminar flow scenario. Turbulent flow at higher flowrates thus results in higher pressure drops compared to laminar flow for the same system.

Figure 5.2 illustrates the difference between laminar flow and turbulent flow in a pipe. In laminar flow, the shape of the fluid profile is parabolic, and flow is smooth. By comparison, in turbulent flow the fluid is vigorously mixed and, because of the mixing and the additional momentum transfer, the fluid profile is much flatter than in laminar flow.

MATHEMATICAL REPRESENTATION OF TRANSPORT PHENOMENA

One of the beautiful things about transport phenomena is that all three types of transport phenomena are closely related and follow similar mathematical relationships. The equations used to describe transport phenomena are called ordinary and partial differential equations. These equations are best covered in advanced college-level mathematics and chemical engineering courses.

---

* Newtonian fluids are common; their viscosity depends only on temperature and composition; their viscosity is not affected by the shear stress caused by convection. By comparison, the viscosity of so-called non-Newtonian fluids varies with the shear stress the fluid is exposed to, and thus non-Newtonian fluids will flow more easily or less easily in different situations, depending on the shear stress the fluid is experiencing; examples of non-Newtonian fluids are catsup and quicksand.

OIL, GAS AND CHEMICALS SECTOR

In his own words
"Combining the fundamentals with continuing education
to ensure optimal process design"

FREEMAN SELF, PE, BECHTEL

Current position:
Bechtel Fellow; Principal Process Engineering, Bechtel Oil, Gas, and Chemicals business unit (Houston, TX)

Education:
- Bachelor's in Chemical Engineering, Georgia Institute of Technology
- MS in Chemical Engineering, Rice University
- MBA, University of Houston

A few career highlights:
- Selected to be a Bechtel Fellow
- Selected to be an AIChE Fellow
- Is a licensed Professional Engineer (PE)

Freeman Self is a design engineer for petroleum-refining, gas-processing, and chemical plants for Bechtel's Oil, Gas, and Chemicals (OG&C) business unit (one of four business units within the company). He has also been elected a Bechtel Fellow—an honor that is reserved for experts who are recognized both within the company and within the industry, who undertake special assignments and mentoring roles within the company. Bechtel, which was founded in 1898, casts a wide shadow in the world of process plant design and construction; it currently employs 55,000 persons and has completed more than 25,000 projects across 160 countries. Bechtel OG&C's primary focus is engineering and construction in the oil, gas, and petrochemical industries.

Stressing the importance of understanding the fundamentals that are taught in the chemical engineering curriculum—fluid flow, heat transfer, and mass transfer, among others—Self says: "As a design engineer, I use many of the fundamental calculations I learned in college, although our work is clearly much more complex, and involves sophisticated software."

The widespread use of calculation and modeling software these days certainly makes the life of the chemical engineer easier on some levels and helps to reduce the likelihood of calculation errors. However, Self notes that constantly evolving software for calculations, modeling, and dynamic simulations is not always foolproof, because the engineer must understand if the model matches the design situation. "Even the underlying calculation methods are constantly changing," he notes. "For instance, we realized at some point that in fluid dynamics, fluid flow calculations needed numerical integration of different forms of the Bernoulli Equation for different types of flow to yield more meaningful results."

To stay on top of these challenges, Self emphasizes the importance of continuing education throughout one's engineering career. "Designs and other information relevant to chemical engineering are always improving, so I devote a lot of time to attending meetings and conferences of the American Institute of Chemical Engineers (AIChE) and the American Petroleum Institute (API), reading professional journals, and purchasing books." Self also regularly publishes papers and presents papers at conferences to contribute his experience to the professional body of engineering knowledge, and he enjoys mentoring young engineers.

DEMAND FOR CHEMICAL ENGINEERING GRADUATES REMAINS STRONG

The future is bright for chemical engineering, and the industry is never stagnant. "People will always want so many things that chemical engineers can help to provide—clean water and clean air, clean environment, safe and abundant food, safety initiatives that will protect plant personnel and assets and the community, and innovative products," says Self. He notes that "facilitating these transformations are constantly evolving computational methods, robust basic research in all areas of chemical engineering, and an unprecedented level of collaboration among different types of engineers and scientists across many different disciplines."

Validating this trend, Self notes, for instance, that the AIChE now has 12 major industrial technology groups in place to foster and facilitate collaboration between industry, academia, and government. "The demarcation between traditional fields is blurring with so many chemical engineering advances that continue to be used by and benefit many different industry sectors."

Just look at the field of biotechnology. "While it was in its infancy just 30 years ago, biotechnology is now widely used in fields as varied as agriculture, pharmacology, the production of chemicals, and more," says Self. He notes that ongoing biotech breakthroughs are encroaching more and more into traditional chemical engineering fields, such as the development of advanced materials, the production of chemicals, drugs, and polymers, liquid-fuels production, packaged foods, and more.

A MAN ON A MISSION

Over the early decades of his career, first at Hudson Engineering, then at Davy McKee, and eventually at Bechtel, Self consistently noticed "that companies typically want people with specific types of expertise to work on specific types of projects—'refinery people for refinery projects' and 'chemical people for chemical projects,' and so on." Sensing that this approach "felt too unstable" and could undermine job stability as certain sectors rose and fell over time with volatile market demand and economic drivers, he made a conscious decision early on to focus on a unifying theme that would always be in demand—process safety.

"I wanted to develop an expertise in something that cross-cut all industry sectors, and when I was starting out I was noticing that process safety practices in general seemed to be very immature and largely qualitative, with lots of 'rules-of-thumb' and institutional practices in use, but not as much consistent scientific strategy or formalized rigor as it needed," says Self. "With an aging workforce and much of the early institutional knowledge eventually being lost, I sensed the opportunity early on to get involved and focus my career on process safety as an essential, ubiquitous type of chemical engineering expertise."

STRIVING FOR EXCELLENCE

Among some of his specific technical achievements, Self has developed new construction materials, new solutions for fluid-flow problems, and practical kinetic models. "In facing any engineering challenge, there are certain attributes that always prove useful: looking at the broad picture, striving to be able to anticipate questions and provide acceptable answers, promoting quality throughout with no shortcuts, and using creativity and understanding of issues to develop solutions," says Self.

"I have had many satisfying experiences and career achievements over many years," says Self, but "two of my most important personal achievements were being elected as a Fellow of the AIChE, and being selected to represent Bechtel as a Bechtel Fellow," says Self, because both signified meaningful recognition among his peers.

THE IMPORTANCE OF EARLY WORK EXPERIENCES AND NETWORKING

Self's father received a chemical engineering degree in 1932, but did not work in the industry after graduation. "I was interested in science, my father always encouraged it, and Georgia Tech required a declared major so I chose chemical engineering," he recalls.

For any chemical engineering student, gaining early work experience is extremely important, for several reasons. "It can help not only when deciding which industries you may want to join, but which industries you may want to avoid," says Self, who adds that by learning and understanding the terms and expectations of employment, students becomes stronger employment candidates upon graduation. While at Georgia Tech, Self was a "co-op student" at Monsanto, so he gained some early experience with the engineering work environment.

Similarly, he notes the importance of networking to build a professional base, saying that it should begin in college and continue throughout your career. "Joining a professional society such as the AIChE, the American Society of Mechanical Engineers (ASME), and others is important to establish a professional home, for mentoring, networking, and continuing education," he recommends, noting that all professional societies have student chapters, and one should continue active involvement after graduation.

Of particular importance, says Self, is the exposure and experience you get from participating and volunteering in professional society events. Such efforts can help build the "soft skills"—such as teamwork and collaboration, public speaking, effective communication, negotiation, leadership, and consensus building—that can later help you to stand out in the workplace. "These groups tend to be very encouraging and create a relatively low-pressure environment that is both inviting and social. Even if you're introverted or shy, it's still a very good place to learn useful professional skills and meet other people in your field," says Self. And the time spent will yield its own dividends over time—building confidence and helping you to build a broad base of friendships throughout the profession, which will be invaluable for long-term career networking and future employment opportunities, he adds.

TECHNICAL VERSUS MANAGERIAL TRACK: KNOW THYSELF

Most chemical engineers will confront a fork in the road as they move through the early years in their careers, at some point being faced with the decision of whether to pursue the proverbial "technical track" (focusing on specialized engineering work) versus the "managerial track" (focusing on managing teams of other technical professionals) at work.

"Most industries do give you a chance to decide which track you want to pursue, but it's really up to the individual to look at his or her strengths and weaknesses, skill sets, personality traits, and proclivities to pursue the most appropriate opportunities," says Self. "I was more interested in remaining on the technical side (even though I have done some managerial stints along the way) and I have continued to focus on that throughout the course of my career," he adds. "You won't necessarily know it early on, but pay attention to your likes and dislikes along the way, and to what other colleagues' career paths look like." ∎

DIVERSE PROCESS OPERATIONS AND INFORMATION MANAGEMENT

In her own words
"Using my flexible chemical engineering degree
in both traditional and non-traditional roles"

VALERIE FLOREZ, PE, CHEMICAL ENGINEER/VOLUNTEER

Current position:
Chemical engineer, temporarily on hiatus following maternity leave; volunteer, working with multiple technical libraries and related organizations

Education:
- BS, Chemical Engineering (Honors), University of California, Berkeley
- MS, Library & Information Sciences, Digital Libraries, Drexel University

A few career highlights:
- Chevron Phillips Chemical Recognition & Achievement Awards—Accomplishments in Support of Process Unit Turnarounds (2001 and 2004)
- Tau Beta Pi Engineering Honor Society (California Alpha Chapter)
- Beta Phi Mu International Library & Information Studies Honor Society (Sigma Chapter)
- Delfino Award for Outstanding Achievement in Chemical Engineering (University of California, Berkeley)
- Society of Women Engineers—TRW Leadership Scholarship
- Is a licensed Professional Engineer (PE)

Valerie Florez holds two degrees—a BS in Chemical Engineering and an MS in Library & Information Sciences. After earning her chemical engineering degree from the University of California, Berkeley, Florez joined Chevron Phillips Chemical Co. in Houston, Texas, as a Process Engineer/Design Engineer. "During my six years there, I focused a lot on process optimization, and I really liked it, she says."

Her work in process design at a world-class chemical company led to an interesting revelation. "Most of my training had been theoretical, and that worked fine in school. But when I started working at Chevron, it quickly became apparent that 'perfect information' is not always available, so that created a shift," she says. "Where do you apply rules-of-thumb? When do you estimate? When are your output or findings good enough? It was not always readily obvious."

"I was lucky in that I worked with some very senior, seasoned engineers at Chevron, and they were able to give me a better sense of what to do when the available information was not perfect," she adds.

During her time at Chevron, Florez developed, maintained, and expanded computer-based tools and models to optimize the facility (including tools used to detect signs of age and wear on mechanical assets), and provided specific recommendations to optimize plant performance (such as an overhaul of purchasing practices and improved equipment monitoring). "Our efforts resulted in improved production forecasting, which ended up saving the company US$3 million per year," she says.

Eventually, she and her husband (who is also a chemical engineer, working for ExxonMobil, and, (profiled on Chapter 1) relocated to Chicago, and Florez took a job at a corn-processing company called Ingredion (formerly Corn Products International). Hired as a process engineer, she characterizes her three years there as being "more of a mechanical engineer than a chemical engineer."

In that capacity, she focused on engineering projects to improve water quality and reduce waste production and water consumption. Through her efforts, she helped the unit in which she worked to reduce waste (and associated costs) by 25 percent and to achieve a 50-percent reduction in clean water consumption. She was also instrumental in developing the monitoring tools that helped to reduce pumping needs on the unit. This helped to not only improve energy efficiency, but reduced water discharge by 42 percent in one year and tank overflow duration by 43 percent.

Specifically, she proposed and developed an idea to reduce water waste by upgrading to a larger, more energy-efficient pump; she also proposed adding an inline pH meter to monitor water quality. Similarly, she developed and executed (from the initial design through bid, fabrication, and installation) a US$250,000 project to improve existing equipment and resolve water contamination issues.

During her tenure there, Florez developed, submitted, and received a permit to execute a US$50,000 project to improve operator safety and operating efficiency; she also oversaw project implementation and the necessary mechanical modifications.

CAPITALIZING ON HER PROJECT MANAGEMENT AND TROUBLESHOOTING SKILLS

After spending three years at Ingredion, Florez and her husband relocated again (as a function of his career advances) to the Fairfax, Virginia area, and she joined Washington, D.C.-based Opower (formerly Positive Energy), a "software-as-a-service (SaaS)" company that was recently purchased by Oracle. She spent four years there, leaving as a Senior Manager in Production Operations.

"The company's software service is sold to utility companies, allowing the utilities to use 'social norming' (a modern-day version of 'keeping up with the Joneses') to help their own customers to reduce energy usage," she explains. While there, Florez was responsible for the production and delivery of millions of so-called Home Energy Reports that were delivered to utility customers.

"Initially, we produced millions of 'Home Energy Reports' that would say 'you're doing better than your neighbor' or 'your neighbor is doing better than you' in terms of energy usage, and eventually those reports moved to a digital format," says Florez. "My role in the company was really unique—I was not a computer programmer. I was really hired because of the many skills I had developed as a chemical engineer." Specifically, she started out as a support engineer and was promoted to Senior Manager of Production Operations at the company.

She notes that the detailed, customer-specific reports produced by the company were printed and sent out by a third-party vendor, and Florez "functioned as the connection between our software engineers and that vendor company."

"I was hired to put processes in place, and the processes had to scale efficiently because the company knew it would be growing over time," says Florez. "A big part of my work there was also to troubleshoot problems—to look for inconsistencies in the reports that would indicate errors in our software or incongruent use of language—and to use continuous-improvement techniques to address issues that arose."

"We eventually hired two more chemical engineers," she says. "I knew they'd do very well in that role," she adds. "Thanks to my chemical engineering training, I could always handle all that was thrown at me, and I knew they could, too."

In this role at Opower, Florez was managing five engineers and coordinators; she led continuous-improvement efforts to first automate many manual aspects of the work, and then spearheaded a major cost-saving transition from one printing technology to another. She was also responsible for meeting mission-critical deadlines.

CHANGING FOCUS DURING MATERNITY LEAVE

During the time Florez and her husband were living in Fairfax, Virginia (when she was working at Opower), she also spent several years (2009–2012) volunteering as an Institutional Exchange Coordinator for the Smithsonian Institution Libraries in Washington, D.C. (assigned to the library for the American Art Museum & National Portrait Gallery). In that capacity, she revived and automated the exchange of materials between institutions, developed and implemented a comprehensive database for the Library Exchange Program, wrote the training manual and trained staff on the use of the new database, and more.

From 2014 to 2015, she also worked as an Independent Contract Cataloger for Mitinet Library Services, working in a virtual format, to improve retrieval rates and discoverability of K-12 materials through in-depth subject analysis using a variety of formats to create robust metadata records for fictional and non-fictional juvenile materials in various languages, including English, Spanish, and French.

Eventually, Florez earned a Master's degree in Library & Information Sciences from Drexel University (Philadelphia, Pennsylvania). "I subconsciously apply chemical engineering principles and process-optimization strategies to everything I do, even when these things don't necessarily seem applicable to others," she says.

Since 2016, Florez has volunteered as a Virtual Reference Librarian for the Oregon State Library. In that capacity, she interacts with patrons from libraries across the United States, Canada, UK, and Australia, providing chat and email support using a program called Question Point 24/7.

REFLECTING ON BEING A FEMALE ENGINEER IN A PROCESS FACILITY

While Florez says the issue did not arise as much at Opower or Ingredion (both of which had a higher ratio of women to men), she recalls being aware of some gender-related issues that arose during her early days at Chevron Phillips Chemical. "When I was working with a variety of engineers in two different locations (a research facility and a pilot plant) during my years at Chevron, I was exposed to many different types of working professionals—engineers, chemists and technicians, office staff, human resources personnel, and more," she says. "The engineers were predominantly male, and considerably older than me, and occasionally issues arose with regard to sexist attitudes."

"Being petite and looking young while working in a predominantly male-dominated environment can create some challenges. In those early days, I was too young to find my voice and really trust my own confidence, especially in the face of attitudes or comments by some of the 'rougher' males. For instance, the males who worked in the field or in a manufacturing facility setting tended to favor raunchy language and humor." Fortunately, she says: "My managers were straightforward, so occasionally when a selective occurrence would arise, they were very supportive."

ADVICE FOR ENGINEERING STUDENTS

Both of her parents are attorneys, but Florez never had any interest in going into law. "I was always good at math and chemistry in high school; when people would ask me what I might want to study in college, I would always say 'I'm going into chemical engineering,'" she recalls. "I always knew that being an engineer would provide a lot of job security and strong salary potential."

"People learn differently—we know this so much better now," says Florez. "We have visual learners, asynchronous learners, people who learn well on their own, people who learn better in a group setting. Some people say, 'I need to see it and hear it,' but I need to take it away. I learn better in an environment away from the fray. The sooner you can learn how you learn, and work from that, the more successful you'll be," she says. "You need to find the best way to do your best."

Reflecting on her college experience, Florez says: "Grades don't always matter as much as you may think they do. Obviously, you need to know the key threshold for grades that may be important for your industry, but don't beat yourself up if you don't get that 4.0. Do the best you can do and stay above the critical thresholds, but don't agonize or drive yourself crazy."

Similarly, she notes that internships are invaluable. "Drexel required a work-study semester and helped students to get a foot in the door," says Florez. "I got my first internship—at Chevron—after the end of sophomore year in college."

"Any introduction will help; and we all need that kind of help, so don't be shy. Turn over every stone, ask everyone you know who they might know, or who they might be able to introduce you to," she says. "Ask for help, push for introductions, that's the only way to make key connections and unearth great opportunities." ∎

CHEMICAL PROCESS RESEARCH & DEVELOPMENT

In his own words
"Chemical engineers: pressed into service
to solve society's most pressing problems"

DAVID DANKWORTH, PhD, ExxonMobil Research and Engineering Co.

Current position:
Distinguished Scientific Advisor, ExxonMobil Research and Engineering Co., Corporate Strategic Research

Education:
- BS, Chemical Engineering, Rice University (Houston, TX)
- PhD, Chemical Engineering, Princeton University (Princeton, NJ), with a year spent studying at Cambridge University

A few career highlights:
- Developed extensive experience in reactor design, project development, and operations related to the deployment of ultra-low-sulfur fuels
- Provided engineering leadership on foreign assignments in Germany, the UK, and Canada
- Served as manager of global process engineering for ExxonMobil refining

David Dankworth has spent the past 26 years working for ExxonMobil Research and Engineering, where he currently holds the title of Distinguished Scientific Advisor for Corporate Strategic Research. He joined the company immediately after completing his PhD in Chemical Engineering, but he has held many different, highly varied positions—15 roles at least, in multiple US and overseas locations—over the past quarter-century. "All of the roles I've held, whether I was doing engineering work, running an organization, developing and managing people, or advancing some specific business objectives, have been closely related to process technology," says Dankworth. "I've always wanted it to be that way."

"I very rarely have asked for specific roles—but I have been very fortunate to work for a company that values talent and looks for it," says Dankworth. "I demonstrated strong capabilities, and I showed promise to do more and different things, so I have been constantly offered new roles that would test those skills."

Following the completion of his PhD in chemical engineering, Dankworth joined the company as a researcher in reactor engineering. Over time, he progressed into management roles, and more recently he has returned to research in his current position, "which is a very senior technical role that serves as liaison between the senior executives and the researchers in the company," he says.

"In my current role, I look more broadly at the various technology needs the company has, and I talk to a lot of experts—for instance, internal scientists and engineers throughout ExxonMobil, university professors, technical experts at the US Deptartment of Energy, and more—and then share the resulting insight to develop technology portfolios that can be considered to address the significant technological challenges at hand."

"I spend a lot of my time encouraging external connections and partnerships with our researchers and engineers, deciding which programs live or die, how we progress through the strategy, how we direct internal research-related resources," he says.

Dankworth has held this Distinguished Scientific Advisor position for a little more than a year. Prior to that he held a series of executive roles, managing groups of engineers in various parts of the company (sometimes as many as 50–100), mostly in the downstream petroleum-refining side, but with work in issues and projects related to chemicals-manufacturing technology, too. "We were the company experts on all of the traditional petroleum-refining unit operations—such as crude distillation, catalytic cracking, and hydro-processing—and we functioned as a center of knowledge and expertise, so we could deploy the resources needed to carry out R&D, project support, and plant support, to address whatever technological issues came up."

ON THE VALUE OF INTERNSHIPS

"When I was in high school, I had the chance to run the Water Department for my home town," says Dankworth " I started working there in the summer after 8th grade and learned how to operate the water-treatment plant, carry out the necessary lab tests, and fix leaks. When the superintendent suddenly retired, I was asked by the Board to take over as a part-time job during my high school years."

He explains: "During that time, I was able to refurbish and restart the idled river-water-treatment plant, add automated level control to the water tower, and replace many old water meters and lines," says Dankworth. "These endeavors were in addition to the monthly billing, collection, and accounting activities I handled."

Through that unusual experience, Dankworth says: "I got a feel for the process and the chemistry and politics behind clean water. And more importantly, based on my hands-on experience, I had a feeling that chemical engineering might be fun."

During his college years, Dankworth had three internships; one was related to civil engineering on a bridge-widening project, where he learned how engineers guide construction projects and ensure a high-quality result.

The other two provided direct chemical engineering experience with Dow Chemical Co. In his first internship with Dow, Dankworth carried out research in a pilot plant that produced polyols as a feedstock to produce polyurethane. "In that polyols-production unit, I learned a lot about practical unit operations and how work gets done in large, complex corporate organizations."

During a second college-era internship with Dow Chemical Co., Dankworth spent the summer "figuring out the kinetics of new catalysts that would remove acetylene from an olefins stream." The experience was very eye-opening for the chemical engineering student. "I could design my own experiments, develop models and put them into Aspen (simulation software), understand reactor engineering and get a much better sense of how it fits into the larger process," he says.

During his PhD program, Dankworth had an opportunity to work at Exxon (prior to its merger with Mobil to become ExxonMobil), in upstream production research. "I was doing really fundamental work measuring the surface tension of decane and water, with different surfactants present," explains Dankworth. "The research was oriented toward improving enhanced oil recovery (EOR) techniques, and we were testing various combinations at the extreme pressure and temperature conditions that would be found underground in petroleum reservoirs. I was measuring the curvature of droplets and the contact angle to measure surface tension in a really fundamental way."

"All of those summer jobs gave me great, invaluable experience and insight into a number of really interesting aspects of the chemical engineering profession," he adds.

CHOOSING CHEMICAL ENGINEERING: AN UNUSUAL PATHWAY

Dankworth attended a very small school (with only 16 students in his graduating class), and he was always interested in (and good at) math and science. To increase the rigor of his studies beyond what his small school was able to offer, he competed intensely in the Texas interscholastic league science competition. This involved reading *Scientific American* magazine and similar journals, developing a solid grounding in chemistry, physics, and biology fundamentals, and ultimately scoring highest on tests at district, regional, and state tournaments. "I earned the gold medal for that test several years in a row, besting students from all but the very biggest regional high schools in Texas," he recalls.

By competing in that contest and making it to the annual state meet, Dankworth was eligible to apply for a Welch Scholarship. "If you get the scholarship, you have to major in Chemistry or Chemical Engineering, and that was the first time I'd heard about chemical engineering," he explains. "I got to understand how chemical engineers are able to be more practical than chemists."

"I didn't know at the time that it was the hardest major!" he says with a laugh. "I went at it more from the perspective that I wanted my scientific education to have really practical applications in the workplace."

"I loved it the whole time," recalls Dankworth. "I was surprised I was able to do the work and I really dug into it. I felt that every year, as I progressed through every aspect of the chemical engineering major, I loved it more—it was just a really good fit."

SEEING THE BIG PICTURE

His advice to current chemical engineering students: "Be really interested in what you're doing and spend the time to absorb all the cool things about it. I feel like some of the kids who are struggling with the major may not really be seeing all the interesting ideas that are there," he says.

Perhaps more importantly, today's chemical engineering students should recognize what an enormous role they will play in society throughout their careers. "Chemical engineers are being asked to take a more active role in the transformation of the world in terms of where we get our energy from and how we create the products we use," says Dankworth. "Important challenges related to sustainability and clean energy are driving the profession. Engineers are supposed to solve the problems, and we're right in the middle of addressing the pressing questions around how we can use science and technology to support 9 billion people on the planet."

KNOW YOUR OPTIONS IN BOTH THE "TECHNICAL TRACK" AND THE "MANAGERIAL TRACK"

"I'm not sure if I really realized this as a student or as a young engineer—and I want readers to understand this: for the chemical process industries, and the oil-and-gas industry in particular, there are usually two developmental pathways for young engineers," says Dankworth. "Some follow a path that leads to greater technical expertise, hands-on engineering work, and value related to that contribution. Others follow a managerial, business-development, or executive pathway that helps them to go on to run a group or eventually the entire company."

"Throughout your career, you will see chemical engineers at the very top of your organization, and you will see them throughout the process engineering and safety aspects of any engineering-related enterprise," says Dankworth. "You will be looked at in both ways, and those opportunities are there for you; so pay attention to your own skills, strengths, and weaknesses, and the opportunities that seem intriguing to you."

"It's very important to build from the foundation of what you learn in college and graduate school, in terms of technical content, teamwork, leadership, and more, and to continue to build on that throughout your career, adding deep knowledge of both technical content and business skills that are relevant to your company and your industry," says Dankworth. "If you do that, you'll succeed and you'll enjoy yourself doing it." ∎

REFERENCES

Moran, Michael J. and Howard N. Shapiro. 2008. *Fundamentals of Engineering Thermodynamics*, 6th Edition. New York: John Wiley and Sons.

Van Ness, Hendrick C. and Michael M. Abbott. 2008. Thermodynamics, Section 4 in *Perry's Chemical Engineers' Handbook*, 8th Edition. Ed. Don W. Green. New York: McGraw-Hill, Inc.

Basic Concepts: Separation Processes and Other Unit Operations

6

C. Judson King (King 1971) explains that *separation processes* are the systematic inverse of mixing processes. Mixing processes increase the uniformity and randomness of the mixed materials. In keeping with the second law of thermodynamics, this results in an increase in entropy of the mixture and its components. Separating the components of a mixture requires the application of thermodynamic work (King 1971).

For example, dissolving salt by mixing it with water creates a resulting mixture of saltwater that has uniform composition. The salt becomes evenly distributed throughout the water, so reversing the process to separate salt and water is problematic. However, using basic principles of chemical engineering, there are at least three options to separate these components again:

- One can apply heat to boil the water off and then condense the steam by cooling.
- One can apply refrigeration and freeze most of the liquid to create pure ice, followed by separation of the ice from the remaining salty brine, and then melting of the ice.
- One can apply pressure to the salt solution and force it through a solid membrane that is permeable to water but impermeable to salt.

All three of these methods have been applied to recover potable water from seawater, but all require energy (King 1971). And from the second law of thermodynamics, all cause a further increase in total entropy.

There are approximately three-dozen available separation processes, each of which is a different unit operation (see definition of *unit operation* in Chapter 2). Separations can be carried out on gas-liquid mixtures, gas-solid mixtures, or liquid-solid mixtures to yield separated constituent streams. A few examples are discussed below.

SEPARATION PROCESS: STRIPPING

Stripping is the evaporative removal of the more volatile component or components in a liquid mixture, and is carried out by contacting the mixture with a gas; this is typically carried out by passing a gas through the liquid mixture, allowing the more volatile component(s) to transfer from the liquid phase into the gas phase and then be removed from the liquid as a vapor that is either captured for reuse or treated (if necessary) and released into the environment.

As an example of stripping, consider the purification of groundwater that has been contaminated with a volatile liquid such as methyl tertiary butyl ether (MTBE) or trichloroethylene.

MTBE was widely used in the latter part of the twentieth century as a fuel additive to increase the octane rating of gasoline. MTBE was also added to increase the oxygen content of the fuel, which decreased the emissions of air pollutants in the exhaust of internal combustion engines. For safety, most automotive gas stations stored gasoline in underground storage tanks (UST). Unfortunately, through the years, hundreds of these underground tanks developed leaks of gasoline containing MTBE, and the leaks were not detected until the plume of contaminated groundwater had reached a well.

Unfortunately, MTBE is sparingly soluble in water, while gasoline is virtually insoluble in water. This means that MTBE easily transferred into the groundwater and imparted a bad taste. Today, thanks to too many instances of such contamination of groundwater, MTBE is no longer

FIGURE 6.1   Shown here are examples of metal, ceramic, and plastic packings in various sizes and shapes that are widely used to fill distillation, stripping, and absorption columns. These packings are used to maximize the amount of contact between the liquid and vapor phases by providing more surface area within the column. The packings provide torturous pathways for the liquid and gaseous streams, thereby promoting turbulence and maximizing the separation of the compounds in the feed to the column. (Source: Shutterstock ID 555205906)

added to gasoline; instead, ethanol is now widely added to gasoline to serve as an oxygenate. And, thanks to the lessons learned (and to reduce the likelihood of leakage), many leaking USTs have been replaced with double-wall USTs that are equipped with leak detection systems between the inner and outer tanks. This practice has helped to drastically reduce the prevalence of leaks from underground storage tanks used to contain potentially hazardous fuels and chemicals.

Air stripping can be used to treat contaminated groundwater to remove the MTBE from the liquid stream by entraining it in an air stream. Specifically, air stripping of groundwater is typically followed by activated carbon adsorption to capture the MTBE and separate it from the air stream, and this has proven to be a successful option. In such a setup, contaminated groundwater is typically pumped to the surface and into the top of a stripping column; air is blown upward from the bottom of the column at as high a rate as possible without entraining the groundwater into the air stream.

The column is usually filled with plastic or metal "particles" about the size of a US quarter and made in a shape that helps to create the maximum interfacial surface area between the air and the water (Figure 6.1). Increased interfacial surface area helps to increase the transfer of MTBE from the downward-flowing water into the upward-flowing air stream.

The MTBE-rich air stream leaving the top of the stripping column is then sent through a bed of activated carbon to separate the MTBE from the air stream by adsorption onto the activated carbon.

INDUSTRY VOICES

"Here's a winning strategy for any student—surround yourself with people who share your values, such as hard work and doing what is right, and distance yourself from those who might jeopardize your success or keep you from working toward the goals you find important."

—Cynthia Mascone,
Chemical Engineering
***Progress* (CEP) Magazine**
and AIChE,
page 95

SEPARATION PROCESS: ADSORPTION

Adsorption is the selective transfer of one or more components from a fluid onto a solid called an adsorbent.

Typical adsorbents include activated carbon, silica gel, alumina, zeolites, and macroreticular polymeric solids. Adsorption takes place because the component (for example, MTBE) in the fluid (for example, air or water) has a strong affinity for the adsorbent material under the conditions of their initial contact. When the adsorbent material has adsorbed its capacity for the target component, it becomes saturated; the saturated adsorbent can then be isolated from the fluid stream—for example, by use of valves on the inlet and outlet of an adsorption column that is filled with adsorbent particles that are a fraction of an inch in diameter (Figure 6.2).

FIGURE 6.2   Granular activated carbon particles, such as those shown here, are widely used as an adsorbent material to selectively remove one or more dissolved liquid or gaseous compounds from a mixture of compounds in a liquid or a gaseous stream. Highly porous activated carbon granules or pellets are valued for their enormous surface area and adsorption capacity for certain molecules that have affinity for it. (Source: Shutterstock ID 697738921)

When activated carbon is saturated with MTBE that has been separated from a fluid, the carbon particles can be safely disposed of by controlled burning in air. Conversely, the activated carbon can be returned to the supplier to be regenerated, using a heat-treatment process (for instance, using steam to desorb or strip off the adsorbed MTBE, followed by burning of the MTBE-rich offgas in air).

Desorption is the reverse of adsorption. It is the process of removing the adsorbed component from the adsorbent material. Desorption can be accomplished either by heating the saturated adsorbent, by pulling a vacuum on the adsorbent, or by displacing the adsorbed component with another compound that is more strongly adsorbed to the adsorbent material.

The ability of many adsorbents to facilitate the removal of an adsorbed component from an adsorbent by a change in conditions (usually temperature and/or pressure) allows for the *regeneration and reuse* of that adsorbent.

> **ASIDE:** The example provided here, of the purification of MTBE-contaminated groundwater by air stripping, is also an example of *countercurrent flow*. The contaminated groundwater enters at the top of the stripping column and flows downward, countercurrent to the stripping air that enters at the base of the column and flows upward. By feeding the MTBE-rich groundwater into the top of the air stripper column, the groundwater contacts the stripping air that has the highest concentration of MTBE with the untreated groundwater that has the highest concentration of MTBE. Because higher concentrations of MTBE in the liquid will yield increased MTBE levels in the stripping gas, this countercurrent flow transfers the most MTBE into the stripping gas.

But the advantages of countercurrent flow do not end there. In the base of the column, incoming stripping air that contains no MTBE is contacted with treated groundwater that has the lowest concentration of MTBE. This gives the maximum removal of MTBE from the groundwater before it leaves the column as treated groundwater. By careful design of the stripping column, and selection of optimal stripping air rates, the resulting MTBE concentrations in the treated groundwater can be reduced to very low targeted levels.

Countercurrent flow also finds many other useful applications in chemical process operations. However, in a few cases, *co-current flow* is preferable. In co-current flow, both phases flow in the same direction.

WHY USE ADSORPTION?

Adsorption is a unit operation that is most effective for decreasing the amount of a chemical in a gas or liquid to a very low concentration by transferring it at a high concentration in a solid adsorbent. One well-known application is the use of silica gel to remove moisture (water) from air.

Silica gel is silicon dioxide that has been processed to form porous particles that have a very large surface area. For example, one typical silica gel has a surface area of 5 acres per ounce or 750 square meters per gram. Water is readily adsorbed on the surface of silica particles, and a monolayer of water molecules on the silica surface can give a water concentration of 30% by weight in silica.

When particles of silica gel are used to dry air that has a high moisture content, one liter of silica gel can remove the water from about 3,000 liters of 100% humid air at 70°F.

Thus, adsorption using silica gel (and other adsorbent materials) can greatly concentrate compounds that exist even at dilute concentrations in fluids; this ability to separate and concentrate the target compound from a more-voluminous fluid stream helps to simplify further processing (for instance, to capture the compound for reuse, or for destruction).

SEPARATION PROCESS: ABSORPTION

Absorption is the selective removal of one or more components of a mixture of gases by contact with a liquid. Note that absorption dissolves the component in a liquid, whereas adsorption adsorbs (or captures) the component on the surface of a solid.

Absorption is also called *scrubbing* and has many applications. Scrubbing of stack exhaust gases that are produced during coal-powered electricity generation is a widespread practice. Coal contains sulfur that is oxidized when coal is burned in a power plant.
Primary reactions:

$$(6.1) \qquad S + O_2 = SO_2 \,(\text{sulfur} + \text{oxygen} = \text{sulfur dioxide})$$

$$(6.2) \qquad SO_2 + H_2O = H_2SO_3 \,(\text{sulfur dioxide} + \text{water} = \text{sulfurous acid})$$

Secondary reactions:

$$(6.3) \qquad SO_2 + \tfrac{1}{2}O_2 = SO_3 \,(\text{sulfur dioxide} + \text{oxygen} = \text{sulfur trioxide})$$

$$(6.4) \qquad SO_3 + H_2O = H_2SO_4 \,(\text{sulfur trioxide} + \text{water} = \text{sulfuric acid})$$

When sulfur oxides in fluegas come into contact with rain, the sulfur oxides are dissolved in the water, producing what is called *acid rain*. During the latter part of the twentieth century, acid rain concentrations in the United States were high enough to begin causing lakes in the northeastern part of the country to become sufficiently acidic to kill fish and other aquatic life.

To counter these severe environmental effects, chemical engineers began designing and installing stack gas scrubbers on coal-fired power plants—a practice that has now become routine.

Stack gases from the firebox of a boiler in a power plant are sent to a scrubber after an initial heat-recovery step and can flow upward through the scrubber. An aqueous solution of a *base* such as slaked lime [$Ca(OH)_2$] is introduced at the top of the scrubber and the liquid fall through the scrubber, contacting the upward-flowing fluegas stream that is rich in sulfur oxides (Figure 6.3).

Here, a *base* is a compound that can neutralize acids by reacting an OH- ion with the hydrogen ion in an acid to produce water. For example, the base sodium hydroxide (NaOH) reacts with the acid hydrochloric acid (HCl) to form sodium chloride (NaCl) and water (H_2O).

FIGURE 6.3   Scrubbers can used to remove unwanted pollutants from an exhaust stream, typically by contacting the exhaust stream with a scrubbing liquid (often water or another appropriate chemical) that absorbs the target pollutants. The pollutants either dissolve or are suspended in the scrubbing fluid, thereby removing them from the stack gas. The scrubbed-clean stack gas is then discharged to the atmosphere. This photo shows steam in a treated fluegas that is being released to the environment from a scrubber tower (with its pollutants removed) at a small industrial plant. (Source: Shutterstock ID 207760669)

The sulfur oxides are removed by chemical reactions with the base material. For example, for scrubbing with slaked lime, Equations 6.5, 6.6, and 6.7 apply:

(6.5) $SO_2 + Ca(OH)_2 = CaSO_3 + H_2O$ (sulfur dioxide + slaked lime = calcium sulfite + water)

(6.6) $CaSO_3 + \frac{1}{2} O_2 = CaSO_4$ (calcium sulfite + oxygen = calcium sulfate)

(6.7) $SO_3 + Ca(OH)_2 + H_2O = CaSO_4\ 2H_2O$ (sulfur trioxide + slaked lime + water = gypsum)

Calcium sulfite is further oxidized to calcium sulfate dihydrate (a product known as gypsum), which precipitates out of the scrubbing liquid as a solid. If sufficiently pure, the gypsum can be captured and used to make wallboard, blackboard markers, and other products.

Scrubbing of power plant fluegas with liquid absorbents such as organic amines is an effective way of removing unwanted carbon dioxide in the fluegas, thereby greatly reducing the greenhouse gas content of the treated fluegas.

Other widely used examples of absorption include the removal of hydrogen sulfide and carbon dioxide from natural gas, and the addition of carbon dioxide to syrup to make carbonated beverages.

SEPARATION PROCESS: EVAPORATION

Evaporation is a process that separates a volatile component of a mixture from non-volatile components by applying heat to the mixture, and/or by pulling a vacuum on the vapor space above the mixture. The overhead vapor product is usually condensed later as a liquid byproduct.

Figure 6.4 is an example of a simple batch evaporation process. Liquid mixture is fed into a column and evaporated by an electric heater in the base of the column. The bottom product,

INDUSTRY VOICES

"Young people need to be fully involved and learn with total passion, and they need to evaluate the technical and financial risks associated with any project, in order to use their engineering expertise to lead and execute projects in the most advantageous way."

—Soni Olufemi Oyekan, Prafis Energy Solutions, page 99

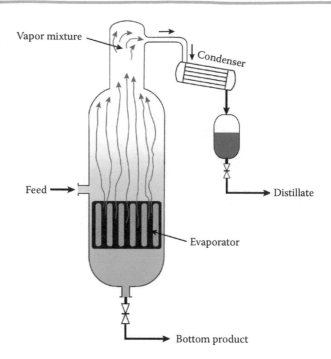

FIGURE 6.4   In this simplified diagram of an evaporation process, a liquid stream that is a mixture of several components enters the evaporator unit and is heated. As the mixture is heated, some of the more volatile components are vaporized. The vapor rises and exits the column through piping at the top of the vessel. The vapor is then condensed to form the liquid condensate, which is rich in volatile components. The remaining bottom liquid stream is removed through the piping at the bottom of the vessel. The bottom product is enriched in the less-volatile components. (Source: Shutterstock ID 510367057)

enriched in the less-volatile components, flows out of the bottom of the column. The overhead vapor stream containing the more volatile components are condensed in an overhead condenser cooled by cooling water.

One common example of evaporation is the concentration of fruit juice by evaporating most of the water content. By pulling a vacuum on the vapor space above a mixture that is stored in a vessel, the mixture boiling point is lowered and the separation of dissolved phases can occur at lower temperatures. Using this approach (to enable separation at lower temperatures) helps to protect the product streams from potential damage that may occur if heating were instead used to promote evaporation needed to separate the phases.

The less-volatile liquid in the base of the evaporator usually contains the desired product. In many cases, when the content of volatile liquid in the bottom of the evaporator is low enough, the non-volatile solid crystallizes out as solid crystals. Sugar crystals are an example product.

SEPARATION PROCESS: DRYING

Where complete removal of volatile liquids is desired, drying can be accomplished by transfer of the moist crystals from the evaporator to a dryer. Several types of dryers exist. The simplest is the tray dryer, in which the moist solids are spread on heated trays, which evaporate the remaining volatile liquid from the solid product. There are several choices of heat source, such as electrically heated, steam-heated, or solar-heated trays, as illustrated by Figure 6.5.

Another type is the spray dryer, in which a concentrated liquid is sprayed into a drying column that is also being fed hot air or another hot gas. The liquid in the falling droplets that are formed in the spray evaporate in the hot gas, leaving behind dry particles that fall to the bottom of the spray dryer.

FIGURE 6.5 The tray dryer is the most common type of dryer. The solid material to be dried is spread across a flat surface (a tray) to maximize the surface area, and the moisture is allowed to evaporate. Sometimes, the process is accelerated by heating the tray surface, or exposing the items to heated air, or by applying a vacuum. Shown here are tomato slices being converted to sun-dried tomatoes in a greenhouse/solar dryer. Industrial dryers typically rely on closed units that heat a material to be dried to accelerate evaporation of the moisture. (Source: Shutterstock ID 654981526)

SEPARATION PROCESS: DISTILLATION

Distillation is a process that uses heat to separate the components of a mixture, based on the different relative volatility characteristics and different boiling temperatures of the individual components in the original liquid mixture. Distillation involves the controlled partitioning of components between gas and liquid phases. After distillation, the liquid mixture of at least two components in the base of the distillation column contains a higher concentration of the less-volatile component(s), while the overhead vapor product contains a higher proportion of the more-volatile component(s).

Continuous distillation occurs when a continuous flow of a liquid mixture is fed to a *distillation column* and is separated to produce a continuous flow of *bottom product* (*a liquid stream*) from the bottom of the column, and a continuous flow of *overhead product* (*a vapor stream*) from the top of the column. The overhead vapor product is then condensed to form a liquid overhead product, or *condensate*. Part of this overhead product is returned to the top of the column as *reflux* (the concept of reflux during distillation is discussed in greater detail below). Heat is provided to the bottom of the distillation column to vaporize part of the liquid. The vapor rising from the base of the column is enriched in the more-volatile component. Consequently, the overhead product is also enriched in the more-volatile component(s).

Reflux is that part of the liquid condensate from the overhead vapor product that is reintroduced to the top of the distillation column.

The return of reflux to a distillation column is a powerful technique to improve the separation of components in the distillation column. When the overhead vapor that is collected from a distillation column is condensed to produce a liquid stream, part of the resulting liquid condensate is returned to the top of the distillation column as reflux and allowed to flow downward through the column. The distillation column is usually filled with inert particles (packing) like those used in a stripping column (see Figure 6.1).

Conversely, the column may be outfitted with horizontal cylindrical plates or "trays" that are spaced at regular intervals (Figures 6.6, 6.7, and 6.8). These trays are designed to allow the rising vapor and falling liquid to mix on each tray, so that the vapor can separate more easily into the vapor rising upward to the tray above, while the liquid will continue falling downward to the tray below. This promotes the countercurrent flow of vapor and liquid inside the column, which increases the concentration of the more-volatile component in the upward-flowing vapor while decreasing the concentration of the more-volatile component in the downward-flowing liquid.

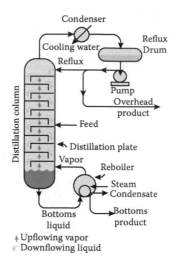

FIGURE 6.6   A distillation column is very useful for separating mixtures of compounds that differ in volatility. A stream containing a liquid mixture of compounds is fed into the middle of the distillation column and flows downward, partially vaporizing as it flows. The more volatile components in the down-flowing mixture are vaporized as the liquid flows downward through the column and contacts the upward-flowing vapor. Vapor is created in the bottom of the column by heating in the reboiler. The resulting vapor stream then flows upwards through plates or packing, which provide intimate mixing with the downward-flowing liquid. The bottom-product stream is enriched in the less-volatile compounds in the original mixture and is sent for use or further processing. In the upper part of the distillation column, vapor continues to rise and exits to a water-cooled condenser, where it is condensed to form the liquid distillate. Usual practice is to return part of the distillate to the top of the column as reflux. This downward-flowing liquid in the part of the column above the mixture feed tray further concentrates more volatile compounds in the upflowing vapor and less-volatile compounds in the downflowing liquid. The overhead product is piped away for use or further processing. (Source: Continuous Binary Fractional Distillation, Wikipedia, contributed by user Mbeychok; used under CC BY-SA 3.0)

When using distillation to separate a mixed-liquid stream, to optimize the separation and increase the concentration of the volatile component in the overhead distillate, engineers can either provide more trays in a taller trayed column, or they can provide a larger volume of granular packing material in a taller packed column (Figure 6.9). This also results in the bottoms product having a still-lower concentration of the volatile component, ensuring a higher level of purity in both the bottom and overhead product streams.

FRACTIONATION OF CRUDE OIL (PETROLEUM)

Crude oil or petroleum liquids are complex mixtures of many organic compounds. Crude oil is so named because the most useful petroleum products are obtained by fractionation or distillation of the starting material (crude oil) to yield different petroleum fractions that have various boiling point ranges and a range of desirable properties. These crude oil fractions are then used to produce a diverse array of intermediate and finished products using various combinations of chemical engineering unit operations.

Figure 6.10 is a highly simplified diagram of a crude oil distillation tower in a petroleum refinery. Crude oil is first heated to more than 400°C and introduced into the bottom of the distillation tower, where it separates into volatile vapors and a hot liquid product containing lubricating oils, petroleum wax, and asphalt. This hot liquid is further processed downstream in a vacuum tower (not shown) to separate and recover the hot liquid components. The volatile vapors rising from the base of the crude tower include products with boiling points ranging from fuel oil to gasoline, LPG (liquefied petroleum gas, which is comprised primarily of propane and butane), and fuel gas. The crude tower contains many trays that facilitate separation by boiling point range of

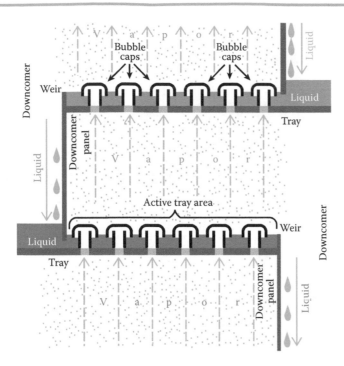

FIGURE 6.7   During the operation of a distillation column that is equipped with trays, the individual trays are designed to allow the rising vapor and falling liquid to mix on each tray. This drawing shows the detail of trays with bubble caps that are used to facilitate mass transfer between rising vapors and downward-flowing liquids. (Source: Bubble Cap Trays, Wikipedia, contributed by user Fabiuccio; used under CC BY-SA 2.5)

FIGURE 6.8   Shown here is a detail of a valve tray used inside in a distillation column. Valve trays are an improvement over bubble caps for contacting rising vapors with downward-flowing liquids and then separating the vapor (which rises to the tray above) and liquid (which flows downward to the tray below through channels called downcomers). (Source: Shutterstock ID 291055505)

the various products. Figure 6.10 shows a typical temperature profile and the products that are removed at various points in the tower. Not shown in Figure 6.10 is the reflux stream from the top of the tower that trickles down through the column, carrying the less-volatile components downward while more-volatile components vaporize and flow counter-currently upward as vapor inside the tower contacts the downward-flowing liquid.

Batch distillation refers to the single liquid charge of a mixture introduced into a distillation column, followed by batch-wise distillation. The distillation process separates the liquid mixture into two phases—an overhead vapor and a residual liquid. The vapor is usually condensed to form a liquid overhead product.

FIGURE 6.9   Distillation column and auxiliary equipment. Distillation is one of the most complex but effective unit operations used to separate mixed fluids into their components. Among other uses, distillation is widely used during petroleum refining, to separate crude petroleum into different petroleum fractions, each of which has its own distinctive properties. These petroleum fractions are then used as feedstocks to produce a diverse array of other intermediate and finished products, using combinations of chemical engineering unit operations. (Source: Shutterstock ID 542377003)

With batch distillation, a certain volume of a liquid mixture is heated in the base of the distillation column, to vaporize part of that liquid stream. The more-volatile component(s) of the liquid vaporize first in this process as heat is applied, while the less-volatile component(s) vaporize less easily. As the stream is heated, separation based on difference in boiling-point temperature produces an overhead vapor stream that has more of the most volatile component(s) in it, and a residual liquid in the bottom of the column that contains a higher proportion of the less-volatile component(s)—hence, the process provides a very efficient way of separating components of the initial mixture, based on differences in the boiling points and volatility of the different components in the stream.

However, if the volatility difference between the less-volatile and the more-volatile components in the mixture is not large enough, the resulting separation may not be as effective. When this is the case, significant amounts of volatile component(s) may remain behind in the residual liquid, and the overhead vapor product may end up containing significant amounts of the less-volatile component(s).

The effectiveness of the separation of the mixture can be improved by providing both reflux and a taller column that contains either more trays or a larger height and volume of internal column-packing materials.

UNIT OPERATION: HEAT TRANSFER

Although we will discuss other separation processes later in this chapter, it is appropriate to discuss the unit operation *heat transfer* because of its importance in evaporation and distillation,

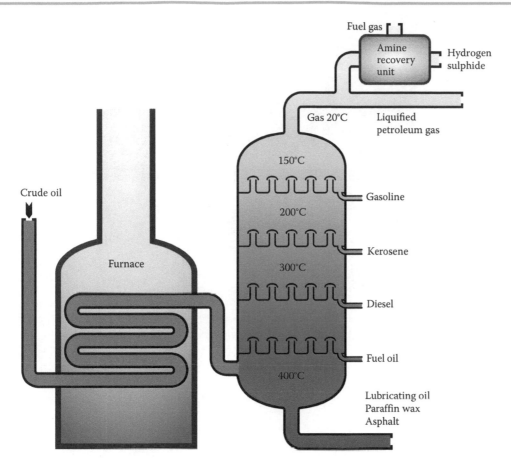

FIGURE 6.10   Fractional distillation of crude oil (often called fractionation) takes advantage of the fact that different constituents within crude petroleum have different boiling points. The process shown here is described in greater detail in the text. (Source: Shutterstock ID 277006853)

as well as many other separation processes. Figure 6.6 shows two types of heat transfer equipment that are used in a continuous distillation process: the *reboiler* and the *condenser*. The *reboiler* is connected to the bottom of the column and is heated (in this case, with steam) to boil the low-volatility liquid found in the bottom of the column. This helps to remove the more-volatile component from the bottoms and send that vapor stream upward for further purification in the column. On the other hand, the *condenser* cools the overhead vapor from the top of the column using cooling water, causing it to condense.

Both the reboiler and the condenser are *heat exchangers*. They both use a utility stream (that is, steam or cooling water) to transfer heat through the wall of a heat exchanger to the process fluid (in the case of distillation, the bottoms liquid or overhead vapor).

The simplest heat exchanger is the *plate-and-frame heat exchanger* (see Figure 6.11). Three metal plates of similar dimensions are separated by impervious inserts around their perimeters. This creates two adjoining compartments, separated from one another by the middle plate. The process fluid to be heated or cooled is introduced to one side of the first compartment, created by two adjoining plates. The utility fluid is introduced to the other compartment on the other side of the central plate. In this way, heat can be transferred across the plate in the middle from the utility side to the process side without contamination of one stream (e.g., the process stream) with the other (e.g., the utility fluid).

The rate of heat transfer can be estimated using the following equation:

(6.8) $$Q = UA(T_{utility} - T_{process}) = \text{heat transferred}$$

FIGURE 6.11   Plate-and-frame heat exchangers rely on an engineered series of plates to significantly increase the surface area inside the unit. Two streams (the process stream and the utility stream) then pass through separate compartments within the heat exchanger and heat is transferred across the plate that divides them from one stream to the other. The amount of heat transfer capacity within such a unit will be directly proportional to the amount of surface area in the unit. The surface area for heat exchange can be adjusted based on how many plates the frame holds. (Source: Shutterstock ID 720651256)

where:

U = overall heat transfer coefficient

A = surface area for heat transfer

$T_{utility}$ = temperature of the utility stream

$T_{process}$ = temperature of the process stream

Often, a single set of three plates for a heat exchanger does not provide enough heat transfer capacity. One common solution is to provide multiple plates, with alternating process compartments and utility compartments. As more plates are added, the amount of heat transfer area, A, increases too, in direct proportion to the number of plates. The stack of plates is held in place by a frame, and such an exchanger is called a *plate-and-frame heat exchanger*.

Another common type of heat exchanger is the *shell-and-tube heat exchanger* (Figure 6.12). These exchangers have a cylindrical bank of parallel tubes that are fitted inside a larger cylindrical shell. Internal barriers within the exchanger isolate the "shell" side from the "tube" side. The process stream flows through one side (for example, flowing along the tube side) of the exchanger, while the utility stream flows through the other side (for example, flowing through the shell side) of the exchanger.

There are many other types of heat exchangers, such as *fin-tube exchangers, spiral exchangers, double-tube exchangers*, and *scraped-surface heat exchangers*, but they will not be discussed here.

Just as countercurrent flow usually gives the best results in liquid–vapor operations like distillation, countercurrent flow in heat exchangers usually yields the best results, as well. For example, when cooling a hot process stream with cooling water, the hot process stream enters the end of the exchanger where warmed cooling water is exiting. The hot process stream transfers the maximum amount of heat to the warmed cooling water. On the other end of the exchanger, once cooled, the formerly hot process stream leaving the exchanger is cooled further by the fresh (and coolest) cooling water entering the exchanger.

FIGURE 6.12   Shell-and-tube heat exchangers are equipped with a cylindrical bank of parallel tubes fitted inside a cylindrical shell; sometimes more than 200 tubes are installed in the exchanger. This design promotes heat exchange between one fluid (which flows inside the tubes, such as a process fluid that is being cooled) and another (such as cooling water, which flows in the vessel but on the outside of the tubes in the shell)

This countercurrent flow maximizes the delta T term ($T_{utility} - T_{process}$) in Equation 6.8, all along the exchanger, increasing the rate of heat transfer and the maximum amount of cooling that can be achieved.

SEPARATION PROCESS: CRYSTALLIZATION

Many compounds that are dissolved in a solvent can form solid crystals when the compound's concentration reaches a level that is above the solubility limit of the compound. This can be achieved either by removing part of the solvent (thereby increasing the concentration of the non-volatile compound in the solution), and/or by decreasing the solubility of the compound (for instance, by reducing the temperature of solution).

Because the internal structure of a crystalline substance is an ordered structural arrangement that is defined by the molecules of that compound, there is often no place within the crystal for contaminants. Therefore, *crystallization* of a compound that is dissolved in a fluid usually results in a *de facto* purification of the solid-form crystalline compound. The purity of the crystals is increased when the solution contains a high concentration of the compound of interest and low concentrations of contaminants. Approximately 90% of active pharmaceutical ingredients (APIs) that are used in life-saving drug therapies are organic crystals and the use of the crystallization process is one of the key unit operations that determine the products' final qualities (Alvarez and Myerson 2010; Wang et al. 2017) when producing APIs that are at the heart of all pharmaceutical products.

SEPARATION PROCESS: ION EXCHANGE

Up to this point, most of the discussion has focused on molecules, but there are also many *ionic compounds* that are also routinely handled by chemical engineers. Many separation processes are used to separate ionic compounds, including some of those already discussed (for example, crystallization or evaporation).

Ion exchange is perhaps the most common separation process for ionic compounds. As an example, consider the softening of water. *Hard water* is caused by the presence of calcium and magnesium salts in raw water. The calcium and magnesium ions form insoluble salts with household soaps, thereby decreasing the cleaning effects of soap, and producing residual scum when soaps and detergents are used—a phenomenon that people typically notice when bathing or washing dishes.

Hard water can be "softened" by passing it through an *ion exchange bed* that contains small polymeric beads of an *ion exchange resin*. These beads are formed from large organic polymeric anions that facilitate the removal of "hardness" ions calcium (Ca^{+2}) and magnesium (Mg^{+2}), substituting sodium (Na^+) in their place. Equations 6.9 and 6.10 show the reactions that occur during the removal of calcium and magnesium ions, and the substitution of sodium ions in their place, by ion exchange in the resin.

(6.9)
$$Ca^{++} + 2Na^+\,resin^- = Ca^{++}(resin^-)_2 + 2Na^+$$

(6.10)
$$Mg^{++} + 2Na^+resin^- = Mg^{++}(resin^-)_2 + 2Na^+$$

Initially, the anionic resin is saturated with cationic sodium ions. Two monovalent (+1 charge) sodium ions are exchanged with each divalent (+2 charge) calcium or magnesium ion. This reaction proceeds rapidly until the resin is saturated with calcium and magnesium ions and the substitution reaction can no longer proceed.

Then, "breakthrough" occurs when the concentrations of calcium and magnesium ions in the effluent from the ion exchange column increase, until the concentration of each reaches the same concentration as the incoming hard water. In the practical application of ion exchange, incoming flow to the ion exchange column is stopped as soon as the beginning of breakthrough is detected. (Breakthrough is confirmed when monitoring devices detect the presence of calcium and magnesium ions in the stream that is allowed to exit the ion exchange unit in the column effluent.)

At this point, the ion exchange resin—now saturated with calcium and magnesium ions—can be replaced with fresh resin. Even better, the resin can be regenerated (and thus made available for reuse) by essentially running the process in reverse; this is carried out by passing a concentrated solution of sodium chloride through the column, to essentially drive out the magnesium and calcium ions that were collected on the resin during the ion exchange process, and replacing them with sodium chloride ions in the ion exchange resin pellets (this allows the resin to be available for the next round of ion exchange). As noted, this regeneration step essentially drives the reactions in Equations 6.9 and 6.10 in reverse, replacing the divalent calcium and magnesium ions in the resin with monovalent sodium ions. Then the regenerated ion exchange column can be reused for softening water in subsequent cycles.

SEPARATION PROCESS: OTHER

There are many other separation processes; King (1971) lists 42 separation processes and groups them into the following three categories:

1. Equilibration separation processes
2. Rate-governed separation processes
3. Mechanical separation processes

(For more information, see King 1971 and McCabe, Smith, and Harriott 2004.)

PROFESSIONAL DEVELOPMENT AND TECHNICAL PUBLISHING

In her own words
"Bringing engineering precision to professional
writing and editing"

CINDY MASCONE, CHEMICAL ENGINEERING PROGRESS (CEP) MAGAZINE, PUBLISHED
BY THE AMERICAN INSTITUTE OF CHEMICAL ENGINEERS (AIChE)

Current position:
Senior Director of Publications at the American Institute of Chemical Engineers (AIChE; New York) and Editor-in-Chief of *Chemical Engineering Progress* (*CEP*) magazine.

Education:
- BS (double major), Chemical Engineering/Engineering and Public Policy

A few career highlights:
- Operated a freelance writing and editing consulting business, called Engineered Writing, while working part-time as Managing Editor of *CEP*
- Was awarded the Distinguished New Engineer Award by the Society of Women Engineers early in her career

"When I was studying chemical engineering as an undergraduate, I was never totally convinced that I wanted to be a chemical engineer, so I've never really held a 'traditional' chemical engineering job," says Cynthia Mascone, who, for almost a decade, has served as the Editor-in-Chief of *Chemical Engineering Progress* (*CEP*) magazine, the flagship monthly publication of the American Institute of Chemical Engineers (AIChE). "My first job after graduating was with the US Environmental Protection Agency (EPA), and that position recognized the value of my undergraduate degree—a double major in Chemical Engineering and Engineering and Public Policy—but this was not a traditional chemical engineering job."

In 1980, while she was working in the EPA's Office of Air Quality Planning and Standards, "Ronald Reagan was elected President, and he had campaigned on a promise to reduce the federal workforce and slash regulations, particularly environmental regulations, so I knew I had to start planning my next move," she says. "I noticed an ad in the back of *Chemical*

Engineering magazine (then a McGraw-Hill publication) for two editor positions, so I interviewed, took a writing and editing test, did well, and joined the masthead as an Assistant Editor in 1982."

"At one point, early in my career as an editor at a technical trade magazine, someone asked me, 'Did you always want to go into journalism?'" she recalls. "I had never really thought about it as being 'journalism'—it was much more about technical writing and editing," she says. Reflecting on the question, though, she recalled how much she had loved working on her high school newspaper. "It turned out, as fate would have it, that I ended up in that profession," she adds. In a twist of fate, when she ran her own freelance writing and editing business, one of her steadiest clients was her son and daughter's high school—where she served as the advisor for the school's student newspaper.

She notes that she developed very solid writing skills during college. "I did a lot of writing as I worked toward my double major. It's hard to teach someone how to write well. You have to just write a lot and learn by doing, and hopefully have some strong professors and mentors who can help you to improve," she says.

MOVING TO AICHE

Mascone left *Chemical Engineering* magazine in 1990 to join AIChE and *CEP*, where, since December 2007, she has served as Editor-in-Chief. *CEP* is the one benefit that every member of AIChE receives; virtually every member gets the magazine in print each month, and 14,000 chemical engineering student members have online access to it.

"In a way, *CEP* is the face of AIChE, and because we are a nonprofit technical association rather than a traditional, commercial publishing venture, we have more flexibility," says Mascone. "While we clearly feel the business pressure to produce a product that advertisers like to support, our bigger mission is to serve our members, and to help them to do their jobs better."

This allows *CEP* to experiment a bit more. For instance, Mascone notes that *CEP* publishes several special sections every year, on such topics as biological engineering, energy, ethics, sustainability, and safety, "which may have relevance to our members and readers, but which may not necessarily have a strong commercial appeal for advertisers of products that are aimed at working chemical engineers."

Mascone says she values "the fact that our Publication Committee and Editorial Advisory Board recognize that the subjects are worthy of editorial coverage, are important to the ongoing advancement of the chemical engineering profession, and will help AIChE members to stay ahead of evolving trends." She notes that she still has a budget and must manage costs, but there is more focus on the inherent intellectual basis of the endeavor, rather than just consideration of the advertising revenue (which is so often the case in commercial publishing).

While Mascone's long career at *CEP* and AIChE have mostly been related to developing relevant editorial materials, she also interacts with other parts of the organization. She routinely attends AIChE meetings—and specialty conferences put on by AIChE—to keep up with developing news and trends, to get ideas for future article topics, and to make contact with speakers who might later convert their presentations into articles for *CEP*.

Similarly, to keep a finger on the pulse of ongoing advances in many industry segments that are related to the chemical process industries, Mascone is also a member of several other professional societies, including the Society of Women Engineers (SWE), the Air & Waste Management Association (AWMA), the American Chemical Society (ACS), and Association Media & Publishing (AM&P).

STRONG EARLY INFLUENCES

Mascone says she couldn't really picture herself working in industry. With her double major in Engineering and Public Policy, she was able to really understand the interface between technology and society. "The combination really helped me to get excited about my studies and look at problems from a much broader perspective. That was key to getting my first job, at the EPA."

Mascone's mother was a math teacher, and her father was a skilled tradesman (a tool- and die-maker for the automotive giant General Motors). "My Dad was a tinkerer who worked with metal all his life and would have studied engineering if he'd had the chance," she says. "And both of my parents always told me they did not want me to feel held back, even though my Dad had not had the same opportunities when he was in high school."

As a female engineering student and early in her career, Mascone took an active role in the SWE. "Not only was my involvement in (Society of Women Engineers) SWE a great network-ing opportunity, but that's where I developed leadership and 'people' skills, especially work-ing with others in groups and on committees, and I gained invaluable career-related skills by interacting with a wide variety of women engineers in different disciplines, industrles, positions, and so on," she says. She recalls that the meetings provided a nice mix of technical and social networking opportunities.

Similarly, Mascone notes that a winning strategy for students is to "surround yourself with people who share your values, such as hard work and doing what is right, and distance your-self from those who might jeopardize your success or keep you from working toward the goals you find most important."

NEW HORIZONS IN CHEMICAL ENGINEERING

As for relevant trends, Mascone notes the tremendous growth in the "bio" area in recent years. "I don't just mean bio in the sense of biomedicine and health-related biological work, but in efforts to use biological organisms as a technological platform," she says. "There's tremendous effort going on in this arena—to produce chemicals using very different, non-traditional routes. We're seeing a lot of research in this area, as well as scaleup efforts to make these new processes viable on a commercial scale."

Since 2016. AIChE has been working with the US Deptartment of Energy (DOE) to lead an effort dubbed "RAPID," which stands for Rapid Advancement in Process Intensification Deployment. The RAPID Manufacturing Institute, part of the government's Manufacturing USA initiative, focuses on using chemical engineering principles and know-how—ranging from separations, catalysis, and transport processes, to kinetics and reaction engineering—to pursue innovation and advances related to modular chemical process intensification (MCPI). The national public-private RAPID consortium involves more than 130 partners (including companies, academic institutions, government laboratories, and government agencies), and has US$70 million in funding from the DOE and another US$70 million-plus in funding from AIChE and its partners. "We're seeing a lot of interest and excitement in this dynamic area," says Mascone.

"Process intensification involves technologies that replace large, expensive, energy-intensive equipment or processes with ones that are smaller, cleaner, less costly, and more efficient, or modules that combine multiple operations into fewer devices or a single apparatus," explains Mascone. Examples include compact heat exchangers, static mixers, microchannel reactors, hybrid separation operations such as reactive distillation and reactive evaporation, and dividing-wall distillation columns, among others. "This is not necessarily a new concept; some of these technologies have been in commercial use for years or even decades, but companies are often timid about being the first to put new technologies to use on a commercial scale." Stakeholders hope that the work of the RAPID consortium will help to advance these principles and deployment in commercial-scale operations.

BRINGING AN ENGINEERING SENSIBILITY TO BEAR ON ALL EFFORTS

Mascone brings her engineering sensibility to all that she does. "One of the things I always emphasize, in my own editorial work and while mentoring my staff and other junior colleagues, is the need for precision in writing. I spend a lot of my time working to clarify the wording of the articles written by our authors who are engineering experts," she adds.

"Instead of asking 'What do you mean by this?' I will ask, 'Do you mean *this*, or do you mean *that*?' That way, the author sees that I do understand what they have written but I am raising very specific questions about something that our readers might possibly misconstrue, based on the way the piece was written."

"Chemical engineers have to be very precise in their calculations and designs, so it's no surprise that, for me, the push for precision in all of my writing and editing would be a primary goal, she says." ■

PETROLEUM REFINING AND CATALYSIS

In his own words
"Applying catalysis expertise to a range
of petroleum-refining challenges"

SONI OLUFEMI OYEKAN, PhD, Prafis Energy Solutions (Richmond, TX)

Current position:
President and Chief Executive Officer (CEO), Prafis Energy Solutions (Richmond, Texas)

Education:
- Bachelor's degree in 1970, Engineering and Applied Sciences (specialization in Chemical Engineering), Yale University (New Haven, CT)
- MS (1972) and PhD (1977) in Chemical Engineering, Carnegie Mellon University (Pittsburgh, PA)

A few career highlights:
- Holds five US patents and nine patents from other countries, for innovations related to the catalytic naphtha process in petroleum refining
- Elected Fellow of the American Institute of Chemical Engineers (AIChE) in 1999
- Honored with the Percy Julian Award in 2009 by the National Organization for the Professional Advancement of Black Chemists and Chemical Engineers (NOBCChE), in recognition of his scientific contributions and achievements in research

Since retiring from Marathon Petroleum Corp. in 2013—after having spent more than 37 years in the petroleum-refining and chemical industries—Soni Oyekan provides consulting services to various industries through his company Prafis Energy Solutions. While he admits to not pushing the consulting business too hard these days—saying that he is "busy enjoying semi-retirement"—most of his clients have included former colleagues or friends who have known him throughout his career and are aware of his experience and technical knowledge in technical issues related to petroleum refining processes and catalysis.

Oyekan came to the United States in 1966 on an academic scholarship offered by the US and his home country, Nigeria, for top-tier Nigerian high school students who had been recognized for their academic excellence to study engineering, medicine, or the sciences. "From my earliest years in Nigeria, I was always interested in academics and had always excelled in my classes, and in Nigeria's National Exams," he says. After high school, he entered a two-year post-high school program and, with outstanding scores in the National Exams, Oyekan applied to the African Scholarship Program of American Universities (ASPAU) group for one of the 15 scholarships that ASPAU had allocated for Nigerian students in 1966. "I passed the written examination and interview and ASPAU sent my paperwork to US universities. I was admitted by Yale University for undergraduate education," he recalls proudly.

Dr. William McClelland of Greenfield, Massachusetts, a 1945 alumnus of Yale, kindly offered to serve as Oyekan's host in the United States. Oyekan says he is "forever indebted to the McClelland family for the love and care that they extended to him between 1966 and 1970 especially, and now."

WORKING IN THE LAB, PRIOR TO UNIVERSITY

Shortly before leaving Nigeria for adventures unknown at Yale, Oyekan held a laboratory job at a small pharmaceutical company (Sterling Winthrop) in Lagos, Nigeria. "I learned more about what it meant to be a working chemist, and I saw what chemical engineers at Sterling Winthrop plant were doing. When I got to Yale, I found that the Chemical Engineering Department, was embedded in the bigger Engineering and Applied Sciences School. The required chemical engineering courses were offered and embedded within the huge school of Engineering and Applied Sciences," he says.

Oyekan went straight on to graduate school at Carnegie Mellon University after graduating from Yale in 1970. "My Master's degree program was a theoretical research study in enhanced oil recovery. After completing that, I knew that I did not want to conduct any more theoretical modeling studies— I really wanted to be more involved with the practical aspects of chemical engineering," he says.

"I was fortunate to have been selected by Professor Anthony L. Dent to conduct my doctoral research and thesis work in heterogeneous catalysis and reactor engineering." Oyekan undertook a catalytic study on the chemisorption, reaction mechanism, and kinetics of the isomerization of methyl cyclobutane over a zinc oxide catalyst.

GETTING A TASTE OF ACADEMIA AS A PROFESSION

To earn extra money during his PhD program at Carnegie Mellon University, Oyekan worked as an instructor at the University of Pittsburgh (and several other community colleges in Pittsburgh), and ended up teaching various courses in chemistry and chemical engineering for the next five years. When he finished up his PhD in catalysis in 1997, he says, "I was really excited to apply the concepts of catalysis and reactor engineering to solve a number of the industrial catalytic problems that were out there in oil refining processes."

Underscoring the versatility and broad reach of catalysis as a concept, he recalls that he had competing job offers with two petroleum-refinery companies, an aluminum company, and a pharmaceuticals company as he was finishing up his PhD. "I needed something I felt that would be challenging, and in the 1970s the petroleum-refining industry was highly focused on research and development (R&D) of oil refining processes. At that time, several of the oil-refining companies were investing a significant amount of money in R&D to stay competitive in oil refining, retail and marketing, and associated business areas," he says. He added: "The leading refining companies had established well-equipped analytical laboratories and had excellent technical experts that provided access to many technologies and capabilities."

Oyekan ultimately joined Exxon Research and Development Company in Baton Rouge, Louisiana* at a senior technical level. "When they described the types of projects I'd be working in [in my job interviews], I thought they were right up my alley," he recalls. "I felt that the Exxon position would offer incredible analytical support, pilot-plant support, high-level support," he adds. "It was a dream job for me, to be able to work in such a high-powered R&D environment straight out of graduate school. It was beautiful and I really enjoyed my three years there."

---

* This was Exxon's name before it merged with Mobil in 1999 to form ExxonMobil.

EARLY WORK IN CATALYSIS LED TO PATENTS

His initial work at Exxon R&D was focused on catalysis for the catalytic naphtha-reforming process and, as Oyekan proudly recalls, "that initial research program led to my first invention (with colleague George Swan) and that led to a high-rhenium platinum (Pt/Re) KX-120 catalyst, and a Pt/Re-combination catalytic system, for naphtha reforming. The result of this initial innovative work led to my first US patent in 1983, and eight related patents issued by other countries," he says. Oyekan eventually earned a total of 14 patents in the course of his professional life.

After being rotated internally, Oyekan completed pioneering, landmark studies on the hydrotreating of unsaturated naphthas such as catalytic cracker and coker naphthas in Baton Rouge. He was attracted by the lure of greater challenges and took a position with leading catalyst manufacturer Engelhard Corp.* as a Group Leader/Section Head, with greater focus on precious metals catalysis in petroleum refining processes.

LEVERAGING OPPORTUNITIES TO SHAPE YOUR CAREER

"The first time Engelhard approached me with a job offer through a 'head hunter' or 'executive search firm' worker, I turned them down," says Oyekan. "I wasn't ready to leave Exxon R&D." Exxon learned of this job offer, and countered by offering him a promotion and a salary increase. However, he notes: "I was able to use that competitive Exxon offer to negotiate for a better salary and benefits at Engelhard. In addition, the personal relationships I was developing with my expected bosses during my interview at Engelhard, led me to ultimately take the Research Section Head position there, and I worked at Engelhard for ten years."

"Some young persons would not have taken the kind of chance that I took, as I had to relocate in 1980 from Baton Rouge, Louisiana, to Piscataway, New Jersey, but I'm grateful I did," he says of this decision-making process.

Ultimately, he ended up working at Engelhard Corp. (in Edison, NJ) and took over as the Group Leader (Research Section Head) of the company's efforts related to petroleum refining and precious-metals-catalyzed petrochemical processes. While at Engelhard, Oyekan earned another US patent for his work on the activation of platinum containing precious metal catalysts. Compared to Exxon at that time and now, Engelhard, and now part of BASF, was a much smaller company, "so there were several opportunities to work in a variety of technical and business areas, and I really enjoyed my time there. I traveled to countries such as Canada, England, Ghana, Mexico, Nigeria, and South Korea. "The assignments associated with interaction with our clients aided in broadening my expertise and experience in oil refining and catalysis, he adds."

After ten years at Engelhard, Oyekan decided to move on to the chemical process side of the aisle, and joined DuPont's Sabine River Works facility in Orange, Texas, where he put his catalysis expertise to work helping the company address air-pollution challenges associated with its Nylon 6-6 process operations (among others). Later, he moved on to Sunoco, to bring his catalysis and reactor engineering expertise to bear on a range of process operations at a half-dozen petroleum refineries. At the time, Sunoco had oil refineries that were located in Marcus Hook and Philadelphia, Pennsylvania; Sarnia, Canada; Toledo, Ohio; Tulsa, Oklahoma; and Yabucoa, Puerto Rico.

After a brief two-year stint at Amoco as one of the three senior process consultants for oil refining processes, and operating out of the Amoco technology center in Naperville, Illinois. While at Amoco, Oyekan, as one of three senior process consultants, had responsibilities for all naphtha processing, hydrocracking, and toluene disproportionation process units, and covered the five oil refineries that were located at Mandan, North Dakota; Salt Lake City, Utah; Texas City, Texas; Whiting, Indiana; and Yorktown, Virginia. He also coordinated the precious-metals management functions for Amoco. BP acquired Amoco in 1999, and Oyekan moved on to work at the Marathon Oil Corporation (MOC).

* Engelhard was acquired in 2006 by BASF.

Marathon Oil Corporation later divested the oil refining, and associated pipeline and oil products marketing businesses, which are now collectively referred to as Marathon Petroleum Corporation (MPC). Oyekan served as the reforming and isomerization technologist, with responsibilities for the naphtha processing units in seven oil refineries. Oyekan worked at MOC and MPC for a total of 14 years and continued to apply his catalysis and reactor engineering expertise in the optimization of catalytic reformers and paraffin-isomerization units. His innovative problem-solving approaches led to him receiving a third U.S. patent in 2014. Two more U.S. patents were issued to Oyekan in 2016, three years after retiring from MPC.

SHARING WISDOM FROM RETIREMENT

Reflecting on all of his varied and challenging positions, Oyekan says: "Young people need to be fully involved and learn with total passion, and they need to learn how to evaluate the technical and financial risks associated with any project, in order to use their engineering expertise to lead and execute projects in the most advantageous way."

On being an African-American engineer, Oyekan recalls that when he first joined Exxon in 1977 after his PhD program, "he was the only black senior technical personnel when he arrived at the Research and Development Laboratories (ERDL) in Baton Rouge, Louisiana." Similarly, when Oyekan arrived in 1980 at Engelhard, he recalls "being the only African-American technical manager."

"Because I had come in from Nigeria, it didn't really faze me. I was fortunate not to have had any of the unwanted heavy baggage of racial discrimination to weigh me down in my early formative and development stages as that could have impacted full realization of my capabilities as an African American. Nonetheless, I'm aware that there are persons who are not receptive to working cooperatively with other people who are different from them," says Oyekan. "The bottom line is that no matter what you look like, you've always got to be as thoroughly prepared as you can, seek great mentors, state your case in technical and business meetings, and hold your ground when the engineering or business case justifies it."

He adds: "As I've worked with people throughout my engineering career, there has been a good proportion of persons who have recognized me for my technical and business contributions—and not my race or the color of my skin.

"I hope that the situation continues to improve globally, and that persons are recognized for their contributions. Their contributions should not be negated or minimized due to color of their skins, small cultural differences, and religious faith differences," says Oyekan. "We should be committed to collectively improving our world, and should work peacefully to preserve the environment and ecosystem for all, and for future generations of people. ■

REFERENCES

Alvarez, A.J. and A.S. Myerson. 2010. *Crystal Growth Design*. 10: 2219.

King, C. Judson. 1971. *Separation Processes*. New York: McGraw-Hill, Inc.

McCabe, Warren L., Julian C. Smith, and Peter Harriott. 2004. *Unit Operations of Chemical Engineering*, 7th Edition. New York: McGraw-Hill, Inc.

Solen, Kenneth A. and John N. Harb. 2012. *Introduction to Chemical Engineering*, 5th Edition. Hoboken, NJ: John Wiley & Sons, Inc.

Wang, Ting et al. 2017. Recent progress of continuous crystallization. *Journal of Industrial and Engineering Chemistry*. 54: 14–29.

Basic Concepts: Dynamics and Control of Chemical Reactions and Processes

7

CHEMICAL AND BIOMOLECULAR SYNTHESIS, CATALYSIS, AND REACTION ENGINEERING

In Chapter 1, the definition of chemical engineering included the transformation of raw materials into useful products by changes in their physical and/or chemical characteristics. Biomolecular engineering plays an analogous role with biomolecules and biochemical transformations. In both disciplines, chemical and/or biochemical reactions are crucial to the creation of many products used by society.

A *chemical reaction* is a process in which one or more chemical species, the reactants, are converted to one or more different chemical species, the products (Folger 1992).

The identity of a *chemical species* is determined by the kind, number, and configuration of that species' atoms (Folger 1992).

A *biochemical reaction* is a chemical reaction that involves one or more different biochemical entities. Examples of biochemical entities are molecules such as antibiotics, sugars, and proteins. Enzymes, the catalytic agents of biochemical change in living organisms, and the living organisms themselves, may also be involved in biochemical reactions and may thus be biochemical entities.

Chemical reactions between chemical species A and B to form chemical species C and D can be represented as follows:

$$(7.1) \qquad A + B = C + D$$

For an **irreversible reaction** (A and B react to form C and D, but almost no C and D react in the reverse reaction to form A and B), Equation 7.2 can represent the **reaction rate** (Solen and Harb 2012; Folger 1992; Harriott 1964).

$$(7.2) \qquad \text{Reaction rate (in units of [moles of } A \text{ / time-volume])} = r_{reaction\,A} = k_r c_A^m c_B^n$$

where:
   k_r = reaction rate constant
   c_A^m = concentration of component A raised to the mth power
   c_B^n = concentration of component B raised to the nth power

Here, the reaction rate is expressed as the quantity (for example, gram moles) of species A reacting in a unit volume (for example, in units of liters) during a unit of time (for example, hours).

A *mole* is a mass of a chemical species that is numerically equal to the molecular weight of that species. For example, water has a molecular weight of 18, so a gram mole is 18 grams of water. A pound mole is 18 pounds of water. And so on.

In most *chemical reactors*, the required volume of a reactor is usually calculated using Equation 7.2 in a mass balance (in units of moles, per Equation 7.2) of the reaction occurring in the reactor.

The sizing and design of chemical reactors will be covered in future courses in chemical engineering. Here, we will just describe some of the more common reactors and explain how they work.

In industrial processes, most chemical reactions are carried out in a reactor. There are many types of reactors, each with advantages for specific types of reactions. The most common types of reactor are the batch reactor, the continuous-flow stirred-tank reactor (CSTR), and the continuous-flow tubular reactor (see Figure 7.1). Each is discussed briefly below.

BATCH REACTORS

A batch reactor is usually a cylindrical tank that is stirred with a motor-driven agitator, with one or more inlets for charging the reactants and any catalysts into the reactor and an outlet for removing the product(s) when the reaction is complete (left side, Figure 7.1).

A *catalyst* is a chemical species that accelerates (catalyzes) a reaction, but is neither consumed nor changed by the reaction.

The reactor may be equipped with a jacket that can provide temperature control in the reactor; for instance, by cooling with a fluid like cooling water or a heat transfer fluid, or by heating with a source of heat, such as steam or hot oil (see Figure 7.2).

Heating and/or cooling can also be provided by pumped recirculation of the reactor contents through an external heat exchanger. When carrying out a batch reaction, to achieve the best results, the reactants and any catalyst are introduced into the reactor in a systematic and pre-planned manner. The contents may be heated to bring the reactants up to a temperature where the desired reaction will occur at an appropriate rate, and to add heat of reaction when conducting an *endothermic reaction* (one that absorbs heat in reaction).

Similarly, cooling is provided to keep the temperature from exceeding safe levels when conducting an *exothermic reaction* (one that releases heat in reaction). Exothermic chemical reactions proceed at a rate that increases exponentially with increasing temperature. If the temperature is too high, the rate of heat generated by a reaction may be so rapid that the reactor undergoes a *runaway reaction*. In a runaway reaction, the rising temperature increases the reaction rate and heat of reaction, which further increases the temperature. There have been many incidents throughout the

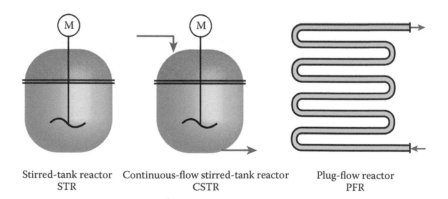

Stirred-tank reactor   Continuous-flow stirred-tank reactor   Plug-flow reactor
         STR   CSTR   PFR

FIGURE 7.1   Three types of chemical reactors. On the left is a stirred-tank reactor that serves as a batch reactor. Reactants are pumped into the tank and then reacted long enough to become the desired product. The reactor is stirred to maintain uniform composition throughout the vessel. Temperature control of the reaction may be achieved by pumping the tank's contents through an external heat exchanger (not shown), or by providing a jacket filled with a flowing heat transfer fluid, such as cooling water. In the middle is a continuous-flow stirred-tank reactor. The main difference between this reactor and the previous batch reactor is the continuous inflow of reactants, and the continuous withdrawal of product. On the right is a plug-flow reactor, which can be a long pipe or a cylindrical vessel with a large length-to-diameter ratio. Fluid flows in one end of the plug-flow reactor, reacting as it flows through, and product is discharged at the other end. (Source: Shutterstock ID 225172513)

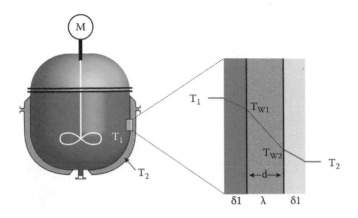

FIGURE 7.2   A typical jacketed reactor on left, with its heat transfer and temperature profile illustrated on the right. The reactor is stirred and kept at a temperature T_1 by cooling with cooling water in the jacket at temperature T_2. (Source: Shutterstock ID 510367087)

chemical process industries and pharmaceutical industry where runaway reactions have caused explosions due to excessively high temperatures and pressures. Batch reactions are particularly susceptible to runaway reactions because the initial reactant concentration and inventory is high, providing energy for a large temperature rise to result in a runaway (uncontrolled) reaction.

When batch reactions are conducted properly, reactants (and a catalyst, if necessary) are added to the reactor. If necessary, the reactor is heated to produce the desired reaction temperature. As the reaction proceeds inside the reactor the reactants are converted to products. When the reactants have been converted, the product mixture is removed and, if necessary, purified using appropriate separation processes.

CONTINUOUS-FLOW STIRRED-TANK REACTORS

Continuous-flow stirred-tank reactors (CSTR) are often similar to batch reactors, but are equipped with one or more pumps to continuously introduce reactants and catalysts (if required). (See the simplified sketch in the center of Figure 7.1.)

The agitator is designed to ensure uniform mixing throughout the reactor. The feed rate of reactants to the reactor is kept low enough so that nearly-complete conversion of the reactants into the desired product occurs inside the reactor. The reaction product is removed continuously at the same rate as the reactant feed into the reactor to keep the mass of fluids in the reactor constant. Temperature is controlled by removing the heat of reaction for exothermic reactions or by adding heat for endothermic reactions. When working properly (i.e., to ensure no runway reaction) the CSTR operates under *steady-state conditions*.

Steady-state refers to a condition in which all points in the process and all variables remain constant. Key variables include reactant and product flow rates and compositions, temperatures, reactor contents, and others.

By contrast, the batch reactor is an example of a process that operates under *unsteady-state conditions*. This signifies that conditions change from the start of the filling of the reactor with reactants to the conclusion of the reaction and the removal of the product.

CONTINUOUS-FLOW TUBULAR REACTORS

Continuous-flow tubular reactors bring reactants (and catalysts, if necessary) into one end of a long tube, and flow at a sufficiently low rate, in a sufficiently long reactor, to allow the reaction to occur during the time interval that the fluid is passing through the reactor (see simplified sketch on the right side of Figure 7.1). The simplest type of continuous-flow tubular reactor is a

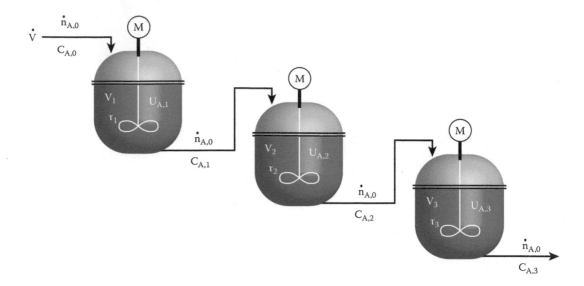

FIGURE 7.3   Shown here are three continuous-flow stirred-tank reactors (CSTRs) in series. When a reaction is carried out in a series of three or more CSTRs, it is similar to the reaction that one would get from a plug-flow reactor. Not shown are the pumps to move the reactants through each tank and then onwards, culminating in pumping the product of the third reactor to downstream processing and use. (Source: Shutterstock ID 510367072)

double-pipe reactor, which is constructed as a smaller tube within a larger tube. In some reactor designs, this is just a small-diameter pipe inside a larger-diameter pipe.

The reactant flows into one end of the smaller tube, and out the other. The heat transfer fluid (to provide either cooling or heating) flows through the annular region between the inner and outer tubes. Fittings on both ends of the tubes serve to isolate the incoming reactant, the outgoing product, and the heat transfer fluid from one another.

The reactant and the heat transfer fluid may flow either counter-currently or co-currently, depending on the reaction and the reactor design. The purpose of the heat transfer fluid is to maintain the reacting fluid at the optimum temperature for the reaction to occur rapidly and efficiently in the reactor. When the tubular reactor is working properly, it operates under steady-state conditions with a defined, predictable temperature and composition profile along its entire length.

Another possible reaction system is a series of two or more continuous stirred tank reactors in series (Figure 7.3). These offer the advantage of different steady-state reaction conditions in each reactor stage. Figure 7.4 is a photo of multistage continuous flow reactors.

Another common type of continuous-flow tubular reactor looks very much like a shell-and-tube heat exchanger (see Figure 6.12). Reactants enter one end of the tube bank and exit the other. The heat transfer fluid enters and leaves the shell side of the reactor, maintaining desired temperatures. The shell-and-tube reactor is frequently used when the reactant fluid is to be contacted with solid catalyst particles to accelerate the reaction. In that case, the tubes are usually filled with the catalyst particles to ensure good contact between the catalyst and the reactant fluid.

REACTORS WITH HETEROGENEOUS CATALYSTS

It is often advantageous to use a catalyst to accelerate reactions and, in some cases, to favor desired reactions over undesired reactions. In some cases, the catalyst is in solution in the reacting fluid, and may become an unwanted impurity that must be separated from the product at the end of the reaction. When the catalyst is dissolved in the reacting fluid, the reaction is said to be *homogeneous*.

In other cases, the catalyst can be bound to small, porous, marble-sized catalyst beads. The presence of many tiny holes in these catalyst support beads creates large surface areas; this helps to promote the desired reactions inside the catalytic reactor. These heterogeneous catalyst particles can be loaded into a tubular reactor and used for an extended period of continuous flow and reaction

FIGURE 7.4   Shown here is a series of cylindrical stirred-tank reactors with dimpled water jackets. Cooling water is circulated through the space between the reactor wall and the surrounding jacket. The dimples help keep the jacket in place and also aid heat transfer to the water flowing though the jacket by promoting turbulence and convective heat transfer. (Source: Shutterstock ID 2660120)

to form product. This system design allows the catalytic reactions to be carried out while helping the product stream to remain uncontaminated by the catalyst. One challenge for heterogeneous reactors is to maintain the optimum temperature within the catalyst bed and catalyst beads. This is because most chemical reactions are either exothermic or endothermic, thus requiring effective heat transfer efforts to ensure desired temperatures within the heterogeneous catalyst particles.

PROCESS DYNAMICS AND CONTROL

In the previous section, the control of temperature in chemical reactors was given as a necessary control function. This section will provide a few basic concepts in the control of chemical process operations.

"*Process control* is maintaining a process output (e.g., a concentration, temperature, or flowrate) within desired specifications by continuously adjusting other variables in the process." (Harriott 1964)

The simplest type of control is *feedback control*. A thermostat that heats a house is an example. The homeowner simply picks the *setpoint*, which is the desired temperature. Then, if the temperature in the house falls below the setpoint temperature, the thermostat (the *temperature controller*) turns on the heater. Once the house temperature rises above the setpoint, the thermostat turns off the heater. The difference between the setpoint temperature and the measured temperature in the house is called the *error*. The temperature controller thus *feeds back* the control signal to the heater to reduce the measured error. The thermostat is called an *on-off controller* because the response to the error is to turn the heater on or off.

There are other ways for a controller to respond to error. For example, a temperature controller with *proportional control* increases the rate of heating in direct proportion to the size of the error in temperature. Proportional control speeds up the response when the temperature error is large, accelerating the return to the setpoint. As the temperature approaches the setpoint, the proportional controller response decreases, reducing the chance of over-heating.

In the early twentieth century, industrial chemical processes were controlled by operators who observed and manually adjusted variables to achieve the desired process outcomes. For example, a holding tank might receive waste acid from an upstream process at a variable flowrate. To control the level in the tank, it is necessary to adjust the flow rate out of the tank. Thus, an operator would stand by the tank and adjust a valve in the tank outlet, as needed, to keep the level

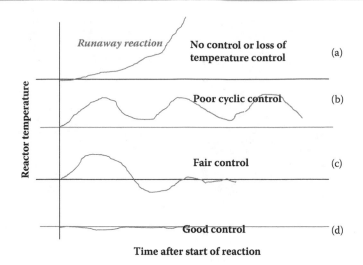

FIGURE 7.5   This figure shows four types of response for control of the temperature of an exothermic batch reaction: (a) shows the reactor temperature running away to unsafely high temperatures due to either inadequate temperature control or total loss of temperature control; (b) shows a cyclic temperature profile due to poor temperature control; (c) shows stable but poorly tuned temperature control that allows large deviations from the desired temperature before eventually reaching the setpoint temperature; (d) shows good temperature control in which the reactor temperature never deviates very much from the setpoint. (Source: Provided by the author)

constant. If the operator stepped away from the tank for a while and the flowrate into the tank increased, the tank might overflow, spilling the acid into the surroundings.

By the mid-twentieth century, mechanical and pneumatic instrumentation and control hardware began to become available that would free the operator's time by providing automatic detection of the process variable, and automatic adjustment of another process variable, to achieve the desired result. For example, a level sensor, a level controller, and an automatic valve could be used to maintain levels in a tank by enabling the controller to adjust the position of the valve in response to changing fluid levels inside the tank.

Dynamic response to changes in process variables must be well understood in order to design a good control system for the process.

Process dynamics refers to the transient or dynamic response of a process to a change in one or more input variables, such as temperature, flowrate, or composition.

Figure 7.5 illustrates the dynamic response of the temperature of a reactor with different control systems. The top figure shows a runaway reaction. The three lower figures show the dynamic response with temperature controls.

Chemical engineers create mathematical models of chemical processes to predict the unsteady state response to changes in input variables. In the last half of the twentieth century and the first years of the twenty-first century, automatic electronic process control systems advanced greatly. Many industrial processes now operate under computer control, using mathematical models to predict the dynamic response of the chemical processes, including the improved response with added controls. Figure 7.6 is a photo of a modern process plant control center.

HAZARDOUS PROPERTIES OF PROCESS CHEMICALS

REACTIVITY

The potential of a runaway reaction—due, for instance, to overheating by an exothermic reaction between two chemicals in a reactor—can result in a fire and/or an explosion; this

FIGURE 7.6   A typical plant control center with multiple screens showing various important aspects of the process and its operating conditions, which typically include temperature, pressure, and composition of various unit operations and intermediate process streams that flow between different unit operations. Operators can assess the state of the process and determine what needs to be done if operation strays outside the preferred operating conditions. The system then adjusts the control variables to make changes in such key variables as the flowrate of cooling water or of reactants; if necessary, the operator and the process control system can initiate a partial or total safe plant shutdown. (Source: Shutterstock ID 356768525)

phenomenon was mentioned earlier in this chapter. *Reactivity* is one potentially hazardous property. Some chemicals can undergo exothermic reactions just by themselves. For example, the monomer chloroprene can be polymerized to make the polymeric *elastomer* (rubber) neoprene. Neoprene is used in many applications where flexibility and chemical resistance are needed. At moderate temperatures, the polymerization reaction of chloroprene can be controlled safely. However, if the temperature rise is too high, chloroprene can undergo a second decomposition reaction that is so vigorous that the pressure rise cannot be vented using conventional emergency *pressure-relief devices* and thus the vessel containing the decomposing chloroprene may explode.

FLAMMABILITY

Flammability is another potentially hazardous chemical property. Flammable liquids are usually characterized by their *flashpoint* (the temperature at which a spark over the liquid ignites a sustained flame). Both gases and vaporized liquids are also characterized by their *lower explosive limit* (LEL)—the concentration of the chemical in air that is flammable, and possibly explosive. For example, methane, the principal chemical in natural gas, has an LEL of 5 percent in air. Methane will not burn in air if the concentration of methane is less than 5 percent (thus, the methane-air mixture is said to be too lean to burn). Similarly, methane will not burn if the methane concentration is greater than 16 percent (UEL, or upper explosive limit). In this case, the methane-air mixture is said to be too rich to burn.

Solids can also exhibit combustibility, as well as explosivity when as dusts dispersed in air. Similarly, when liquids are dispersed as mists in air, their flammable range is extended.

TOXICITY

Toxicity is another hazardous property of chemicals. For example, the toxicity by inhalation of carbon monoxide gas depends on the duration of exposure. The National Research Council has established the following *Emergency Exposure Guidance Levels* (EEGLs): 1,500 ppm for 10 minutes exposure, 800 ppm for 30 minutes exposure, 400 ppm for 60 minutes exposure, and 50 ppm for 24 hours exposure.

Toxicity also depends on mode of contact. In addition to inhalation, the three other main routes of exposure are eye, ingestion, and skin. The US National Institute for Occupational Safety and Health (NIOSH) lists the following categories of chemical agents that are hazardous by toxicity:

- Biotoxins
- Blister agents
- Incapacitating agents
- Lung-damaging agents
- Nerve agents
- Riot control/tear agents
- Systemic agents
- Vomiting agents

CORROSIVITY

Many chemicals are corrosive or irritating to body tissue. They pose a risk to exposure by skin, eye, inhalation, and ingestion. Most corrosives are either acids or bases. Common acids include the following:

- Hydrochloric acid
- Sulfuric acid
- Nitric acid
- Chromic acid
- Acetic acid
- Hydrofluoric acid
- Concentrated solutions of certain weak acids

Common bases include sodium hydroxide and potassium hydroxide.

Due to the corrosive and often toxic nature of many acids and other hazardous substances, protecting humans and the environment from exposure to them is the primary task for many chemical engineers working in industrial settings. Such protection is provided through the properly engineered design, operation, control, and maintenance of any chemical process facility, by personal protective equipment (PPE), and by appropriate product design.

It is also important to note that many of the hundreds of chemicals of commerce pose only limited hazards during manufacture, and that most are very safe when used in products. Toward that end, as discussed in the next chapter, it is essential that all potential hazards be identified early, as processes are being designed to produce specific chemicals and products so that the risks to people and to the environment be carefully minimized and monitored throughout the entire process and product life cycles.

SAFETY-RELATED INSTRUMENTATION AND CONTROL

In her own words
"Deploying the latest risk-analysis techniques
and system design to ensure plant safety"

ANGELA SUMMERS, PhD, PE, SIS-TECH

Current position:
President and Founder, automation and safety consultancy SIS-TECH

Education:
- BS, Chemical Engineering, Mississippi State University
- MS, Chemical Engineering, Clemson University
- PhD, Chemical Engineering, University of Alabama

A few career highlights:
- Authored the Safety Instrumented Systems chapter in Section 23, Process Safety, of the 9th Edition of *Perry's Chemical Engineering Handbook.*
- Authored Safety Instrumented Systems chapter in the 2018 edition of the *Kirk-Othmer Encyclopedia of Chemical Technology.*
- Lead editor, *Guidelines for Safe Automation of Chemical Processes, 2nd Edition*, Center for Chemical Process Safety, American Institute of Chemical Engineers, 2017.
- Named as an Engineering Fellow of the University of Alabama, and a Centennial Fellow in its Chemical Engineering Department (an award given to the "100 most-notable graduates in the first 100 years of the program")
- Named as a Fellow of the International Society of Automation (ISA), the American Institute of Chemical Engineers (AIChE), and the Center for Chemical Process Safety (CCPS)
- Holds eight patents in the United States and 28 internationally
- Is a licensed Professional Engineer (PE)

Angela Summers has more than 30 years of industry experience, covering instrumentation and control, process design, and environmental pollution control. She currently operates three businesses, two related to automation and one unrelated (discussed below), all of which utilize her chemical engineering expertise. The first is a consulting and engineering business (SIS-TECH), which specializes in instrumentation and controls for safety-related applications throughout the chemical process industries. The second is a manufacturing business that produces hardware devices and related software tools to support various safety-related engineering functions. The third is a winery that she and her husband own in Oregon.

Summers started up SIS-TECH more than 15 years ago "with $1,000 and a desire to stop process safety incidents," she says. Today, the company has more than 50 employees and contractors and has become a respected provider of state-based control systems with special expertise in process safety and critical control applications.

SIS-TECH specializes in the assessment, design, programming, operation, maintenance, and long-term support of instrumented safeguards that are used to prevent catastrophic releases in the chemical process industries. "Our engineering clients—ranging from both onshore and offshore oil-and-gas processing facilities, chemical and petrochemical plants, petroleum refineries, pulp-and-paper mills, pharmaceutical operations, and more—need experts to come in and evaluate the risk they have identified and to verify that they have put in place the right monitoring devices, control systems, operating procedures, and protocols to meet the prevailing standards and codes, to reduce risk, and to protect plant personnel, plant assets, and the environment," says Summers. "Proper instrumentation and process control play an integral role in reducing risk and managing process hazards, and what we bring to the table is expertise in risk assessment, automation system design, and maintenance troubleshooting."

"We started out with consulting, mainly with smaller companies that needed to retrofit existing systems," she says. "Most of the safety-related systems that were available were too high-end, too costly and too complex for these smaller companies, so we developed a product to fit that niche and, over time, we've grown to expand our offerings."

A WEALTH OF RELATED AND UNRELATED EXPERIENCES

"I've had an interesting set of detours along the way that have helped to shape how my career has evolved over time," says Summers. "I started out as a process engineer at Ethyl Corp., but I always had a lot of interest in environmental issues, as both my Master's degree and PhD work were focused on waste cleanup."

In the 1990s, Summers spent about a decade working for an environmental company (ENSR), where she developed patented processes related to waste cleanup, and for a safety controller manufacturer (Triconex), conducting risk assessments for clients.

"The interesting thing about having a chemical engineering degree is that you have so many options open, because the concepts you learn—thermodynamics, kinetics, fluid flow, and mass flow—are essentially the fundamental rules about how everything works," says Summers.

"By the time I was in my 30s, I had process engineering experience, had provided support for a physical plant, had carried out novel experiments in a lab, and had held a position with a strong instrumentation-and-controls focus," she says. "That might seem like a real 'ricochet' career trajectory for a lot of people, but for a chemical engineer it is just a good example of how flexibly the degree can be applied in different job environments. I am a stronger and more knowledgeable person today because of all of that experience."

One of the biggest surprises she had when moving from school into the workplace was the amount of learning she still had to do. "So much of what you learn about math and simulation modeling in school does not apply directly to real-world situations," says Summers. "A lot of my colleagues who graduated at the same time as me had a hard time bridging the gap between the intangible simulation of numbers and the real-life aspects of the plant. We are so trained in the equations and simulation rules that it can be hard to embrace all of the real-world complexity."

"Until they get hands-on experience in a process plant environment, young engineers won't really have an adequate perspective on how equation-based modeling must be assessed against the vagaries of real-world conditions," says Summers. "Many young engineers don't really have a firm understanding of what the simulation is showing; they know that changing some input will change the output, but such complex modeling has real-world implications for process control and safety that cannot be approached like they are some type of video game."

BEING A TRAILBLAZER AS A FEMALE ENGINEER

During her first summer job, working at International Paper Company in the early 1980s, Summers says she "witnessed a lot of sexist attitudes and behaviors that would not be tolerated in the workplace today." She recalls that "it was pretty severe and I could see it was taking its toll on some of the other women who were working there, who were taking it personally and having their confidence shaken by it. I'd like to presume that many of the off-color or sexist remarks that happen in male-dominated workplaces are made for shock value—but it certainly doesn't make it right."

"From very early on in my college program, I was often the only female in the math and physics classes I was taking," says Summers. "I just became very comfortable being me, and not what people needed me to be by way of some stereotype. I don't remember ever thinking 'I can't do this.' I have a positive attitude and just always believed I could, and I have counseled other women over time on this issue."

"When I was finishing up school, it was harder for many female students to chart a path because there weren't many role models in the industries we were pursuing," she says. "It's gratifying for today's students to see female chemical engineers in all sorts of roles, and to see female executives managing entire companies, operations, and engineering and environmental groups," she adds.

CHEMICAL ENGINEERING IN THE VINEYARD

Driven by a passion for Pinot Noir wine, Summers and her husband bought Willamette Valley property in 2004 and planted a vineyard to produce wine grapes, naming it Saffron Fields Vineyard. "We fell in love with the wine and the entire biological/chemical engineering process, so we produced our first bottling in 2010, and opened a tasting room in 2013," she says. Currently producing about 2,500 cases per year, Summers and her husband intend to grow production to 5,000 cases per year in the years to come.

The entire process involves many biochemical engineering principles, basic unit operations, and instrumentation. These include the processes that are used to cull the leaves and remove the stems from the grapes, the diaphragms used to press the juice from the grapes, the fermentation process used to produce the wine, and the wine aging, fluid-transfer, and bottling operations. "The whole process involves successive stages of cooling, filtration, the use of pumps, valves and more, often carried out in small-batch operations," she notes. "It is perhaps less complex and a bit more straightforward than a typical chemical process plant, but, of course, the output of this system is far more tasty!" Summers says with a smile.

ON CHOOSING CHEMICAL ENGINEERING

"Early on, I considered studying psychology because I had some intrinsic interest in the subject, but my mother urged me to consider chemical engineering because I was good at math and chemistry," says Summers. "In the early years, piling physics on top of my other courses was not particularly enjoyable for me, and the early semesters surely wore me down a bit. But then I got into thermodynamics and it all started to come together and make physical sense beyond the equations. I took every thermodynamics course that I could after that, even in graduate school—I just fell in love with chemical engineering."

For students and young engineers to be successful, they should be mindful of these three basic sets of rules, says Summers.

1. You need to show up, be ready to act, do something, and speak up.
2. Conversely, you need to know when to keep your mouth shut and to listen, to identify barriers and challenges going forward; don't assume you know everything before you've had a chance to hear from everybody else.
3. Once you see a path forward, act on it, write something, stand up and speak, and then do it every single time.

"Don't be surprised if your voice isn't heard if you don't show up consistently," she adds. "Get involved in the AIChE and be as active as possible in groups and committees. All that involvement will bring its own rewards."

Living her advice of "be ready to act," Summers initiated the formation of the Instrument Reliability Network at Texas A&M University in 2012, to share historical information and lessons learned. This network seeks to improve industry safety, maximize asset performance, and reduce maintenance costs through better lifecycle management of instrumentation and controls applied in the chemical process industries. To date, she has published more than 70 technical papers, technical reports, and chapters for engineering handbooks and industrial society reference materials.

As for the future, Summers predicts that there will be a lot more focus on human factors and how they impact plant operations. "We've already had a lot of focus on ergonomics (in terms of back injuries and carpal tunnel syndrome in workers), but as an industry we're going to be focusing more on the operator interfaces," says Summers. "This will likely play out in the form of better ways to display information. Chemical engineers are going to get more involved in that space, in terms of asking and answering questions related to enhancing operator effectiveness during abnormal operation and emergency operations, pushing the current design paradigm for the human-machine interface." She adds: "We're being challenged to think more holistically about both the operator and his or her interfaces to the process equipment, machinery, and control systems."

BRINGING HER EXPERTISE TO BEAR THROUGHOUT THE PROFESSION

In recognition of her contribution to the development of standards and processes for process safety management (particularly ISA 84 and IEC 61511), especially in the use of automation to prevent loss events in the process industries, Summers has received the distinction of being named Fellow for several industry organizations: the International Society of Automation (ISA), the American Institute of Chemical Engineers (AIChE), and the Center for Chemical Process Safety (CCPS). Summers recently completed editing the 2nd edition of the CCPS book *Guidelines for Safe Automation of Chemical Processes*. Similarly, for many years, Summers has served as an Adjunct Professor, teaching college courses each spring at the University of Houston–Clear Lake, within the Environmental Management Department.

She reflects with pride on one particularly important feather in her proverbial cap: "I remember with great pride buying my own copy of the 6th Edition of *Perry's Chemical Engineering Handbook* when I was an undergraduate, as I knew it would be the encyclopedia of the career I had chosen. When I had the opportunity to write the chapter on Safety Instrumented Systems for the 8th Edition of the book, it meant so much to me." ■

ACADEMIA—CHEMICAL ENGINEERING

In his own words
"Memo to students: education is the only thing they
can never take away from you"

IRVIN OSBORNE-LEE, PhD, Roy G. Perry College of Engineering, Prairie View A&M University

Current position:
Professor and Head of the Chemical Engineering Deptartment, Roy G. Perry College of
Engineering, Prairie View A&M University (Prairie View, TX)

Education:
BSc, MSc, and PhD in Chemical Engineering, The University of Texas at Austin

A few career highlights:
- Director of the Gulf Coast Authority (GCA)
- Director of the American Institute of Chemical Engineers (AIChE)
- AIChE Fellow
- Associate Director of the Center for Environmentally Beneficial Catalysis (CEBC)

For nearly 20 years, Dr. Irvin Osborne-Lee has been Professor of Chemical Engineering and
Head of the Chemical Engineering Department at the Roy G. Perry College of Engineering
at Prairie View A&M University. He jokes that he "once *worked as* a chemical engineer... but
now I *produce* chemical engineers as my final end product." But all joking aside, Osborne-Lee
says he views this role—helping to support, share, and inform the chemical engineers of
tomorrow—as a "really powerful responsibility."

Osborne-Lee reflects proudly on his "upward career spiral," noting that he "went from
being an 'inner city kid' growing up in Houston, to a young man who earned three chemical engineering degrees (BS, MS, and PhD) at The University of Texas at Austin, to spending
13 years at the Chemical Technology Division of the US Government's Oak Ridge National
Laboratory (ORNL) in Tennessee, to joining Prairie View A&M University in Texas, where he
now serves as Professor of Chemical Engineering and Head of the Department (for nearly
20 years)."

"My PhD program in chemical engineering prepared me well to carry out research projects and to be an independent researcher—to really take a problem and begin to understand
it, formulate a path to solve the problem, carry out the work, and communicate effectively (in
both written and oral format) about the whole process," he says. As he was pondering his first

professional role, he decided he did not want to work in a process plant or other industrial facility. Instead, he joined the federal government's ORNL as a researcher.

"My original work at ORNL was not directly related to my PhD dissertation, but that work gave me all of the competencies I needed," he says. His early work at ORNL focused on developing the theory and the models needed to predict the re-suspension of aerosols that would be produced in the event of a potential nuclear power plant accident scenario, in order to predict how much radiation would be released in the event of a nuclear meltdown. "Models had already been developed to predict how much radiation would be released and deposited; my work was to investigate what would happen if gusts of wind began to move the deposited particles around again," he explains.

Interestingly, Osborne-Lee met his wife, Anita Osborne-Lee (whose profile is featured in Chapter 9 of this book), when both were students at the University of Texas at Austin (she was an undergraduate when he was a graduate student). After they graduated, both Irvin and Anita joined ORNL; she joined ORNL's Environmental Science Division to work on environmental monitoring and data collection about six months before he completed his PhD "We both liked it very much and stayed for the next 13-or-so years," he says.

"One of the things I particularly enjoyed about my work at ORNL was that the federal lab functioned as a pseudo-industrial setting, but still had a strong primary focus on fundamental R&D—because the facility is owned by the US Department of Energy, but is operated by an industrial contractor [at the time Osborne-Lee was there this was Martin Marietta, but it is now Lockheed Martin]," he says. "So technically I was a corporate employee, and I gained terrific industry perspective, but I got a lot of perspective about how government-owned research was carried out."

Eventually, when Anita left ORNL to pursue a law degree, the entire Osborne-Lee family relocated from Tennessee to Texas, where she enrolled in law school.

A CAREER IN ACADEMIA BECKONED

"When I was still a graduate student, I had a premonition that my career would ultimately end up in academia," says Osborne-Lee. So when he sought to leave ORNL and relocate to Texas, he pursued his current position at Prairie View A&M University.

"One of the most awesome things about being a chemical engineer is all of the options you have," says Osborne-Lee. "I've seen my chemical engineering students land high-paying, challenging, and intellectually satisfying positions in the chemicals and pharmaceutical industries, in the electronics and semiconductor space, in both the traditional (fossil fuels) and alternative-energy industries and nuclear engineering fields, in the agricultural industry, in bioengineering and biotechnology-based processes, and in non-traditional fields, such as law and financial services."

He notes that the flexibility of this inherently versatile degree is particularly useful when economic conditions create ups and downs in specific industrial sectors. For instance, Osborne-Lee notes that a downturn in oil prices may cause a temporary slowdown in hiring in the traditional fossil fuels space (petroleum and natural gas exploration and refining).

But the flip side of that coin is that when oil prices fall, it creates a boon for all of the industrial segments that rely heavily on petroleum and petroleum-derived products as the key feedstock, such as the chemicals and plastics industries, among others. "This helps chemical engineers to weather certain economic ups and downs better than other types of engineers," he says.

"Chemical engineering is an endlessly flexible, adaptable, and dynamic field of study that is forever launching entire new technologies," he adds. "These include advances in nanotechnology, pharmaceutical breakthroughs, computer and molecular modeling, and simulation of chemical processes (in terms of both theoretical efforts and software development)."

EXCELLENT MENTORS CAN CHANGE LIVES

Osborne-Lee was an "excellent academic performer" in high school—someone who always liked math and science. He recalls that his Chemistry and Calculus teachers "happened to be good buddies, and they conspired to direct me and motivate me to go to college."

"I was pretty clueless and had not really thought much about college, but these two teachers got me thinking about it and figuring out a way to make it happen," he says. In fact, he recalls that "those two teachers would give me a pass to miss class so I could meet with the Guidance Counselor and fill out every single scholarship application that was available."

"The counselor said 'you should write on these scholarship applications what your likely major will be,' so I poured over books and discovered chemical engineering, where my strength in math and chemistry would be useful," says Osborne-Lee. "I saw that the salaries associated with that major were strong. I was also attracted by the fact that, at that time, chemical engineering tended to be one of the smallest of the engineering disciplines, so majoring in it would make me more employable."

RECALLING HIS COLLEGE EXPERIENCE

Recalling his own college experience—and speaking from the perspective of working as a chemical engineering professor—Osborne-Lee says "all of the engineering disciplines provide a challenging course of study, but for me, I really enjoyed being challenged like that." He adds that study of chemical engineering requires students to take some classes that cross over into other engineering fields (such as statics and dynamics, electrical circuits, and strength of materials). "I really didn't enjoy these courses as much—so that reaffirmed my love of chemical engineering," he says.

And, importantly, he adds: "Chemical engineering has always allowed me to really understand how things work, even when I'm looking into other disciplines, and math is the language by which you can make use of scientific principles and bring them to bear on any situation," he says.

ON MENTORING HIS STUDENTS TODAY

"I see students struggling in many cases, both as an instructor and as head of the department," he says, adding that he typically uses two strategies when trying to counsel students who are having a hard time. "When students approach me saying they are unhappy with the way their performance or grades are going, I remind them that 'if nothing changes, then nothing changes.'"

"Any student who is struggling with a challenging set of courses or concepts must reflect and make the personal changes needed to succeed, whether that means re-evaluating priorities around how they are spending their time, or seeking out all resources available. Such support may come in the form of attending office hours with professors or teaching assistants, or seeking such useful resources as the Schaum's "Solved Problems" series of books, which provide an endless array of solved problems for engineering students to use outside the classroom.

"Sometimes students have to declare a 'red alert' and tell their friends and partners that they must simply go underground for a while, to regain control over their coursework and meet their deadlines," he says. "That must always be the top priority when pursuing any type of degree, and everything else must follow after that."

"There are simply times when you must put friendships and all other distractions on hold and do what's necessary," says Osborne-Lee. "All students have it within themselves to make the right choices and priorities, and to develop winning study habits—no student should ever fail because they did not put in sufficient time and effort to succeed." ∎

REFERENCES

CCPS (Center for Chemical Process Safety), 2016. *Guidelines for Safe Automation of Chemical Processes*, 2nd Edition. New York: American Institute of Chemical Engineers.

Folger, H. Scott. 1992. *Elements of Chemical Reaction Engineering*, 2nd Edition. Englewood Cliffs, NJ: Prentice Hall, Inc.

Harriott, Peter. 1964. *Process Control*. New York: McGraw-Hill, Inc.

Solen, Kenneth A. and John N. Harb. 2012. *Introduction to Chemical Engineering*, 5th Edition. Hoboken, NJ: John Wiley & Sons, Inc.

Summers, Angela. 2008. Process Safety, Chapter 23 in *Perry's Chemical Engineers' Handbook*, 8th Edition. Ed. Don W. Green. New York: McGraw-Hill, Inc.

Basic Concepts: Creation and Stewardship of the Sustainable Process Plant

8

THE SUSTAINABLE PROCESS PLANT

Most chemical processes are made up of several interconnected unit operations and reactions. For example, a *chemical manufacturing process* might include the following steps:

- Preparation and purification of raw materials to produce the reactant(s)
- Chemical or biochemical reactions of the reactant to form product
- Separation and purification of the reaction product(s) from byproducts and any remaining unconverted reactants
- Recycling of any unconverted reactant back to the reaction step
- Sale or disposal of any byproducts
- Conversion of product into the final form
- Sale of product
- Closure and decommissioning of the plant
- Any necessary remediation of the plant site (this should be limited to removal of the infrastructure if proper environmental practices have been followed during plant operation)

Many choices need to be made in the selection of the processing steps and their arrangement, size, and operating conditions. Traditionally, in the mid-twentieth century, this selection was done by the chemical process development engineer, based on his/her experience, along with other project team members, such as project managers, chemists, mechanical engineers, electrical and instrument engineers, civil engineers, and designers. In the past, environmental and process safety considerations were often given less attention than the capital and operating costs of the plant and the product.

However, creating and operating a sustainable process plant—one that has a minimal impact on the environment and provides maximum safety assurances for workers and the community—is now the primary goal in the twenty-first century. The sustainable process plant has the following features.

- The entire production process (starting from the acquisition of raw materials through to the creation of the facility) and the products produced by the plant have minimal adverse environmental impact.
- The potential occupational and process safety risks posed by the entire production process have been eliminated, minimized, or controlled to an as-low-as-reasonably-practical level (ALARP), and in accordance with government regulations, relevant regulatory permits, and industry-accepted practice.

PROCESS SIMULATION AND PROCESS SYNTHESIS

When designing a new process plant, *process simulation* (using computer) software is often indispensable. In the past, process simulators could predict the detailed steady-state performance of individual unit operations. Simulators have advanced to the point that the simulators can now also accurately predict the unsteady-state response of simpler chemical processes, given the

FIGURE 8.1   Today, most engineering drawings are created electronically on a computer model of the process and plant. However, large printouts of drawings can be convenient for editing and for checking in the field. (Source: Shutterstock ID 732482194)

configuration of the unit operations and reaction step(s) in the process. For given flowrates, temperatures, pressures, and compositions for the incoming process feeds, these simulators can often predict the required size of the process equipment needed to achieve desired process results.

These simulators are also used to optimize process designs, usually by simulating the optimization of one unit operation at a time (for instance, by evaluating the impact of changing various process variables using a simulated model of the process). Today, equation-oriented solutions of an entire flowsheet are also possible (Figure 8.1).

Finally, simulators can be used to train operators in the plant's dynamic response to potential process upsets, and to the plant's response to operators' interventions.

In the last several decades, chemical engineers have developed an advanced methodology called *process synthesis* or *process integration* to further aid the process engineer in optimizing the arrangement of the processing steps. These techniques are created to go beyond simulating just one unit operation at a time; instead, process synthesis or process integration aims to take advantage of complex interactions between separate unit operations within the process plant. For example, when there are many heat exchangers in a process, process integration efforts can lead to large energy savings by coupling those process streams that need cooling with those that need heating. In this way, fewer utilities like cooling water and steam are required, reducing operating costs.

PROCESS AND PLANT DESIGN, CONSTRUCTION, AND STARTUP

A key step in the creation of a new process plant is the process design by the chemical engineer, and the design of the plant surrounding the process by the other engineers and designers on the design project team (Figure 8.2). As discussed in the next chapter, process simulation and process synthesis are just part of the early steps in the creation of a new process plant.

Before process development begins, product and process research takes place. The most appropriate technology is selected to carry out the desired chemical process operations, either using existing technology, or using a new process and/or product that was created during the process research step. The process design developed by the process engineer predicts the required size and operating conditions of the equipment in the process. This preliminary process design provides the core concepts for the entire plant design team. During this *front-end engineering and design* (FEED) phase, engineers with different types of specialization, working together, each has a role to play; several examples are given below.

FIGURE 8.2   Teamwork is essential in most of the industry segments using chemical and biomolecular engineering principles. Chemical and biomolecular engineers often work with other engineers, designers, and technicians when conceptualizing and designing a new process plant, and moving through construction, start-up, and operation. Each team member brings special expertise and responsibility for specific parts of the overall design project. (Source: Shutterstock ID 525920641)

- Mechanical engineers create preliminary mechanical designs and cost estimates of vessels like columns, reactors, and heat exchangers, as well as rotating equipment like pumps and compressors, using equipment types and sizes provided by the process engineer.
- Piping engineers and designers, using data from the process engineer, design the complex piping systems that connect the vessels and equipment throughout the process.
- Instrumentation engineers and designers create preliminary designs of instrumentation and control systems.
- Electrical engineers and designers create preliminary designs of motors and the plant's electrical distribution system.
- Civil/structural engineers make preliminary designs of structures and foundations needed to house and support the new process equipment and piping.
- Cost estimators make a preliminary estimate of the total cost of the new facility.
- Scheulders develop a schedule that includes the duration and staffing of tasks and activities in the project.

At the completion of the FEED phase, management must decide whether to commit additional funds to carry out a detailed design and a more accurate cost estimate for the new process plant.

If the economic analysis and market conditions suggest that the plant will be profitable and safe to build and operate, *detailed engineering* begins. At this point, the process engineer makes final changes to the process design and provides detailed specifications of process equipment and piping to the other engineers of the plant design team; the design team then prepares purchase orders for new vessels, equipment, and materials. For vessels and equipment that take a long time to create, preliminary purchase orders may be issued earlier in the process, to shorten the time required to complete the plant.

When the plant design is virtually complete, the team updates the cost estimate and the schedule for the plant construction. The owner company then decides whether to move toward initiating construction, or to postpone or cancel the project.

If the decision is to go forward with construction, the design is finalized and vessels, equipment, and materials such as piping, structural steel, and concrete deliveries are built into a detailed project schedule (Figure 8.3).

After completion of the design and construction phases, the new plant is first *commissioned* by carefully inspecting and testing all aspects to confirm that construction adheres to design, that all equipment is tested and operating properly, that normal and emergency operating procedures

INDUSTRY VOICES

"As chemical engineers, you will develop an instinctive understanding of the way things behave, and get a clear picture in your head to really understand what's going on in different spaces. Such problem-solving skills have broad applicability—not just in solving broad scientific and engineering challenges, but in managing large teams and complex projects."

—Susan Weiher, Global Foundries, page 131

FIGURE 8.3   Chemical and biomolecular engineers are primarily involved with the design, start-up, and operation of plant facilities rather than their actual construction. However, they have important roles in site selection and in assuring that the newly constructed facility is built according to specifications. Shown here is an engineer and a construction foreman with construction drawings at a new infrastructure site. (Source: Shutterstock ID 590113328)

(written by chemical engineers) are in place, and that the operating and maintenance teams have been assembled and trained in the operation and maintenance of the new plant. Once the plant has been thoroughly checked and tested, commissioning is complete, and ongoing management responsibilities are handed over from the design-and-construction team to the operation-and-maintenance team.

Once these teams are in place, *plant start-up* begins, with a dedicated *start-up team* working closely with the operating team. The start-up team often includes members of the project team and the commissioning team, and usually includes chemical process, mechanical, and instrumentation engineers.

GREEN ENGINEERING AND PROCESS SAFETY MANAGEMENT (PSM)

Two important chemical engineering disciplines that have developed during the past several decades are **green engineering** and **process safety management** (PSM). These two disciplines have some of the same objectives and methods, but they have largely evolved separately. In both cases, these disciplines should be integrated into chemical engineering practice, throughout the entire lifecycle of any project—from research and development to process and plant design, through plant operations and enterprise management, and all in other areas of chemical and biomolecular engineering practice.

Green engineering is the environmentally conscious design of chemical processes (Allen and Shonnard 2002).

Allen and Shonnard have published a classic textbook that outlines the basic principles of green engineering. That textbook is now freely available on the web from the US Environmental Protection Agency (EPA 2017).

Human activities are having measurable effects on the biosphere, so it is essential for chemical and biomolecular engineers to have a fundamental understanding of these environmental effects; examples are listed in Table 8.1.

Chemical and biomolecular engineers, who are educated in the principles that create and operate these products and processes, are uniquely qualified to make the engineering and operational changes needed to minimize adverse environmental effects, through improved product and process design.

A traditional environmental engineering approach to minimizing environmental effects has been to install treatment systems on the so-called "end-of-the-pipe," to capture or destroy liquid, gaseous, and solid pollutants prior to their being discharged to the atmosphere. One example is

TABLE 8.1   Examples of Adverse Environmental Effects of the Process Industries

1. Destruction of high-altitude atmospheric ozone during the second half of the twentieth century by manufactured refrigerants like chlorofluorocarbons, which led to the formation of the "ozone hole" above Antarctica. In response, harmful refrigerants have been phased out and replaced by more environmentally friendly alternative options. Chemists have devised new refrigerants that do not harm the ozone layer, and chemical engineers have designed and built the plants to manufacture them.

2. Acid rain from the combustion of high-sulfur coal in electric power plants, which causes surface waters (such as ponds, lakes, and rivers) to become highly acidic and threaten wildlife, among other consequences. Now sulfur emissions produced during the combustion of fossil fuels are routinely removed by fluegas scrubbers, and an ongoing transition from coal to cleaner power sources, such as wind and solar electricity production, is helping to prevent the formation of acid rain.

3. Accumulation of the pesticide Dichlorodiphenyltrichloroethane (DDT) in the fatty tissues of people and wildlife, causing effects such as thinning of egg shells of birds and other harmful consequences. DDT, although a very effective pesticide, is no longer manufactured or used in the US and many other parts of the world because of its harmful environmental effects.

Sources: Allen, David T. and David R. Shonnard, *Green Engineering – Environmentally Conscious Design of Chemical Processes,* Upper Saddle River, NJ: Prentice Hall PTR, Prentice-Hall, Inc, 2002.

the installation of scrubbers on the stacks of coal-fired power plants to capture the sulfur oxides that cause acid rain. Another example is the installation of biological wastewater-treatment systems on the liquid discharge from a chemical plant. A third example is the installation of pump-and-treat systems to remove MTBE from contaminated groundwater.

While end-of-the-pipe treatment approaches are necessary, many argue that a better approach is *pollution prevention,* which involves efforts to modify the process to eliminate the source of the pollutant(s) as far upstream as possible, so that fewer contaminants (and/or lower waste volumes) are produced overall. One example of pollution prevention is the replacement of coal with natural gas. It is common practice for petroleum companies that recover natural gas to use chemical engineering processes to remove sulfur compounds. Natural gas provides a cleaner fuel option to generate electricity, as it produces fewer greenhouse gases and no sulfur oxides—not to mention no coal ash—compared to coal.

A second example of pollution prevention relates to the topic of fuel oxygenate Methyl tertiary butyl ether (MTBE) being soluble in water. As noted in Chapter 6, using MTBE as a gasoline additive can pollute water and groundwater sources that may come in contact with gasoline released from leaking underground storage tanks. Replacing MTBE with ethanol (which is readily biodegradable) as an alternative oxygenate in gasoline provides a way of eliminating that source of pollution.

Historically, traditional chemical engineering focused on the properties of the product that met only the narrow needs of the consumer, and those chemical and thermo-physical properties that were necessary to design the manufacturing process. Now, to anticipate potential adverse environmental effects of a potential new product, the potential environmental impact of the product must be evaluated throughout its entire lifecycle—from the raw materials and the original creation of the product to its final fate.

An environmental screening review for a new product might evaluate the properties listed in Table 8.2 (Allen and Shonnard 2002) for the product, as well as for any byproducts and wastes.

The choice of the best product also depends on its manufacturing process. The choice of process to make the product should not depend solely on cost; rather, the process must also be evaluated and designed for in terms of sustainability and environmental safety. In short, it is essential to view the entire lifecycle of a potential product, and the process and resources used to produce it.

PROCESS SAFETY MANAGEMENT

While the emphasis in green engineering and green chemistry is on protection of the environment, the focus of process safety management (PSM) is on prevention of hazardous events in chemical processes and the elimination, control, and minimization of their adverse effects.

Process safety management (PSM) is a management system that is focused on the prevention of, preparedness for, mitigation of, response to, and restoration from catastrophic releases of chemicals or energy from a process associated with a facility. (Center for Chemical Process Safety 2007a)

TABLE 8.2 Chemical Properties Needed to Perform Environmental Risk Screenings

Environmental Process	Relevant Properties
Dispersion and fate	• Volatility, density, melting point, water solubility, effectiveness of wastewater treatment
Persistence in the environment	• Atmospheric oxidation rate, aqueous hydrolysis rate, photolysis rate, rate of microbial degradation and adsorption
Uptake by organisms	• Volatility, lipophilicity, molecular size, degradation rate in organism
Human uptake	• Transport across dermal layers, transport rate across lung membrane, degradation rates within the human body
Toxicity and other health effects	• Dose-response relationships

Source: Allen, David T. and David R. Shonnard, Green Engineering – Environmentally Conscious Design of Chemical Processes, Upper Saddle River, NJ: Prentice Hall PTR, Prentice-Hall, Inc., available for free from EPA website, www.epa.gov/green -engineering/green-engineering-environmentally-conscious-design-chemical-processes-text-book (accessed August 20, 2017), 2002.

Unfortunately, through the years, there have been far too many hazardous events in the process industries. The worst industrial accident ever was the release of highly toxic methyl isocyanate (MIC) at a Union Carbide plant in Bhopal, India in 1984. At least 4,000 people were killed and about 100,000 were injured. Although chemical engineers had been holding routine professional meetings to focus on loss-prevention methods since the 1960s, the devastating events that happened at Bhopal were so much larger and more catastrophic than previous hazardous incidents, so it became clear that a dramatic advance in process safety was essential.

In 1985, again in response to the events that took place at Bhopal, the US chemical industry and the American Institute of Chemical Engineers (AIChE) created the AIChE Center for Chemical Process Safety (CCPS), which still plays a leading role in the chemical engineering community's efforts to ensure process and environmental safety (Center for Chemical Process Safety 2007a; Sanders 2014). Most chemical companies are now members of CCPS and, through CCPS activities, several hundred chemical process safety engineers from industry work together throughout the year to advance process safety.

To date, CCPS has published more than 100 guideline books on various aspects of process safety and holds training courses routinely. The AIChE and CCPS organize international conferences, such as the annual Global Congress on Process Safety (this book's co-author Victor Edwards chaired the 2013 CCPS Global Congress in San Antonio, Texas). Well over 1,000 industry professionals attend the global CCPS congress every year to learn new methods and practices, and to foster greater professional understanding and collaboration to improve and optimize process safety.

THE ESSENCE OF PROCESS SAFETY PLANNING

The following points are key to achieving a safe chemical process throughout the lifecycle of a process plant.

1. *Create a process safety culture* throughout the organization that places high value on process safety.
2. *Identify hazards* by conducting systematic hazard reviews of the process by a team that includes, for example, the project manager, a chemical process safety engineer (who facilitates the review), a chemical process engineer, a chemical technology engineer, a mechanical engineer, an instrumentation and control engineer, a piping designer, a maintenance specialist, and a production specialist. Once hazards are identified, capture them in a *hazard register* that describes the origin and elimination, control, or mitigation of the hazard. Continuously update the hazard register for the life of the facility to be certain that all hazards are properly addressed.
3. *Identify practical ways to eliminate, control, manage, or mitigate hazards* to achieve an acceptable level of risk.

4. *Make these needed changes* to the process.
5. *Practice excellent process safety management* throughout the process plant and the entire enterprise.

One promising new approach to the *inherently safer design* of chemical processes incorporates the following concepts (Center for Chemical Process Safety 2007b).

1. Minimize the inventory of hazardous substances that are stored onsite. (The idea, according to chemical engineer Trevor Kletz, widely considered to be the creator of many inherently safer design principles, is that "what you don't have [onsite] can't leak.")
2. Substitute less-hazardous substances for more-hazardous substances where possible.
3. Make process operating conditions less severe (e.g., use lower process pressures and process temperatures closer to ambient where possible).
4. Simplify complex processes, operating procedures, and control systems.

OPERATIONAL DISCIPLINE AND PROCESS SAFETY CULTURE

Many process plants transform large amounts of chemicals and/or fuels, sometimes at high pressures and temperatures. Thus, incidents in process plants have the potential to cause serious human, economic, and/or environmental harm.

The practice of process safety management (PSM) has developed to minimize the risks of incidents. Human error is often the ultimate cause of failures, including failures during the operation of process plants. Ensuring a strong process safety culture in an organization means that all people involved value process safety highly and are committed to ensuring that process safety principles are followed to minimize incidents (Figure 8.4). This commitment to process safety minimizes the risk of incidents. A concept that is closely related to strong process safety culture is *operational discipline* (OD). OD means complying with a set of well-thought-out and well-defined processes, and consistently executing them correctly.

Operational discipline (OD) is the deeply rooted dedication and commitment by every member of an organization to carry out each task the right way every time. Expressed more succinctly, OD means "Everyone Do It Right, Every Time."

FIGURE 8.4   This operator is checking process operating conditions at an offshore oil and gas platform. Thanks to today's automation and control technologies, critical process information is typically available in the control room and remote locations. Nonetheless, it is also important to check operation conditions of the process machinery in the field. (Source: Shutterstock ID 482493412)

CHEMICAL OPERATIONS AND SAFETY

In his own words
"Supporting safety excellence
throughout a network of industrial facilities"

OTIS SHELTON, PRAXAIR, INC./UNION CARBIDE CORP. (RETIRED)

Education:
- BS, Chemical Engineering, University of Houston
- MS, Chemical Engineering, University of Houston

A few career highlights:
- Fellow of the AIChE (an honor reserved for AIChE members who have proven to be strong contributing members to the institute, society, and the chemical engineering profession)
- Served as AIChE President in 2014, on the AIChE Board of Directors from 2000 until 2002, and as AIChE National Secretary from 2004 until 2006
- Graduated Salutatorian of his high-school class (a distinction that led to his being offered his first professional summer job, working for Union Carbide)
- Life member of the National Society of Black Engineers (NSBE), and served on NSBE's Advisory Board for 20 years (1986–2006)
- Grand Knight of the St Edwards Knights of Columbus Council (2012–2013)
- Recipient of the AIChE's Minority Affairs Committee Pioneers of Diversity Award (2016)
- Recipient of the NSBE Chair's Council Award for Outstanding and Dedicated Service to the NSBE (2015)

Before retiring in 2012, after a decades-long career in chemical engineering, Otis Shelton spent 20 years working in the Safety & Environmental Services Deptartment of Praxair, Inc., one of the largest industrial gases companies in the world, which has a global reach spanning North and South America, Asia, and Europe.

"I was hired by Praxair to develop and direct its Corporate Compliance and Operational Assessment Program. Through this program, we managed and conducted coordinated in-depth assessments, using representative sampling techniques and through in-depth protocols, at Praxair's industrial gas facilities worldwide," says Shelton. "We developed a methodology to assess a range of important safety-related programs, including personnel safety, environmental safety, distribution, industrial hygiene, process safety, product safety, and emergency planning."

Prior to his employment with Praxair, Shelton spent 25 years working for Union Carbide (the parent company to Praxair, before it was spun off as a stand-alone company in 1992). He served in a variety of roles on the Texas Gulf Coast, and at worldwide headquarters in New York City and Danbury, Connecticut.

PIONEERING A NEW APPROACH TO PROCESS AND PLANT SAFETY

"In the early 1980s, Union Carbide developed a first-of-its-kind assessment program with the support of Arthur D. Little," says Shelton. "The company's goal was to ensure that we had thorough, second-to-none safety audit protocols in place to verify compliance. The company sought to verify that its safety programs, design safety, and practices ensured safe operations and prevented major safety events." He notes that, in the first year of the program, "we conducted more than 200 assessments throughout the company."

In 1992, when Praxair (then the industrial gases division of Union Carbide) was spun off to form its own company, Shelton was asked to join Praxair, to develop and run its Corporate Compliance and Operational Assessment program. "I helped to design the overall program, recruit in-company resources, and develop protocols so that we could conduct safety assessments at a variety of Praxair facilities around the world, including air-separation plants, packaged-gas facilities, so-called HyCO (hydrogen/carbon monoxide) plants, electronic gas plants, and surface-technologies facilities, to name a few," he says. "We had to have a thorough audit process and representative sampling to ensure that our assessments of safety were accurate."

"The Compliance and Operational Assessment Program was one of many initiatives undertaken by Praxair safety-program management and Praxair senior management to improve its safety performance," he adds.

BUILDING ACCOUNTABILITY USING A "REPORT CARD" APPROACH

"At the time, Praxair's CEO wanted the assessments to be graded, and a draft assessment report to be communicated to plant management shortly after it was conducted," explains Shelton. "Our audit process allowed us to leave a detailed report, with grades assigned for each safety issue or area that was reviewed." This process allowed plant management to promptly address and correct deficiencies, and to communicate the assessment results to Praxair senior management to ensure follow-up."

"Typically, an assessment's scope includes compliance-related issues—such as environmental performance, product safety, personnel safety, industrial hygiene, management systems, and transportation-related issues—but we needed to evaluate operational safety by looking at process operations and control strategies, as well." says Shelton. "Achieving these ambitious objectives was a challenge, and I'm happy we accomplished it."

"We created a disciplined way to identify key issues that needed to be corrected in a plant's safety-management system and, importantly, we ensured that line management and senior management remained both informed and involved throughout the process," he adds. "In doing so, we were able to help ensure that consistent safety practices were being followed throughout Praxair's numerous facilities all over the world."

TEAMWORK: A KEY ELEMENT TO THIS APPROACH

"The key was to develop a team approach," says Shelton. "We involved not just safety professionals on the assessment teams, but also the plant managers and Technical Center experts who had strong technical knowledgeable of operational processes," says Shelton. "The feedback we got was that the teams we engaged were really knowledgeable in the underlying processes and practices—they were not just following a checklist."

He adds that this team approach—which was novel at the time Shelton and his team were developing this program—was only possible with the strong leadership and support the program received from Praxair's senior management worldwide. "With this support from the top, we were able to send independent, knowledgeable teams to carry out these assessments in Praxair facilities worldwide."

"Over the past ten-or-so years, the concept of defining and implementing consistent operational discipline has evolved, but I believe we may have been early pioneers in helping to ensure such operational discipline was, in fact, implemented in our operating facilities," he adds. "We were doing really groundbreaking work to improve safety and accountability through assessing operational discipline in our plants."

"Similarly, all of our safety-audit findings were tracked on computers. The plant had to review the assessment finding, develop corrective action, and prioritize how to deal with them or fix them," says Shelton. "We developed a history that allowed us to look at past assessments as we were conducting the latest one, so there was accountability and a formal set of checks and balances."

GETTING A FOOT IN THE DOOR

Shelton was fortunate to have spent five summers working at Union Carbide's Seadrift and Texas City facilities during college and graduate school. "Not only did that help me to pay for my education, but I got a lot of very good experience working at the plant—starting as a laborer, then in engineering, then in research-and-development, and then in production," he says.

After high school, Shelton presumed he would major in chemistry, a subject he really enjoyed. "However, during the first two summers I worked at the Union Carbide plant in Seadrift, Texas, I got a first-hand opportunity to see the type of work that both chemists and chemical engineers perform in the plant, and to see how their work impacted the manufacturing goals and results," says Shelton. "Based on my summer experience, I selected the chemical engineering profession, because it allowed me to use chemistry and put it to use in many practical applications. I knew what to expect and I loved it."

APPRECIATING THE RESPONSIBILITY HE WAS GIVEN EARLY ON

In one of his early summer assignments, Shelton was working in a unit that produced low-density polyethylene. "This production process had to achieve pressures of 37,000 psi," he explains. "I worked on both the pumping systems that were needed to create the required pressures in the reactors that were used to convert ethylene to polyethylene in a very sensitive reaction."

During this assignment, Shelton says: "I leveraged my multi-disciplined chemical engineering curriculum and knowledge to review, evaluate, and improve the design and operation of pneumatic pumping systems, and high-pressure packings and seals, to improve mechanical reliability."

Similarly, Shelton recalls that during some of his on-the-job orientation shift work he was tasked with walking through the process plant, drawing lines and valves and process diagrams to help familiarize himself with the process. "So when it came time for me to be the process engineer for the unit, I knew how all the components of the process were connected," he says.

Shelton says he felt fortunate to have had a variety of opportunities to use his chemical engineering degree throughout his tenure with Union Carbide. In addition to his work in the company's low-density polyethylene production unit, and a vinyls production unit, Shelton also spent several years as a purchasing agent for process equipment, and as a coordinator of operations at Union Carbide's Texas City Marine Terminal. Later, he worked as a corporate human resources recruiter for engineers, helping with recruitment efforts at universities located in the southwest.

"As a process engineer with responsibility for a part of the manufacturing production unit, I was responsible for daily operation of the unit to meet the business and customer needs, as well as improving the process operations," he explains. "The plant operates 24/7, except for scheduled maintenance outages. As such, I provided daily operating instructions for operators, and if issues developed after the day shift, I was on call at any time to provide technical advice on actions to take to address plant operating issues."

"A key part of your responsibility as a process engineer is to improve the operating process. To do this, I worked closely with Union Carbide experts from the Technical Center, initiated projects to make improvements in, for example, percent operating time, product quality, maintenance costs, technical process improvements, and safety (process safety, personnel safety, industrial hygiene, environmental safety, and more)."

Reflecting on his chemical engineering education, he says that his many engineering classes at university "prepared me to understand the technology for operating processes, to evaluate risks and hazards, to understand the potential consequences of deviations, and how to apply chemical engineering fundamentals to create and implement plant designs and improve processes." Examples include evaluating pump performance, installing new catalyst additions for the reactors, implementing many different projects to upgrade manufacturing performance, evaluating refining column performance, evaluating the performance of dryers, preparing recipes for daily production runs, and troubleshooting production and product quality issues.

Meanwhile, as a Purchasing Agent in the company's Gulf Coast Consolidated Purchasing group, Shelton says he could leverage his manufacturing knowledge, send out bids, and evaluate proposals for process equipment such as heat exchangers, pumps, compressors, distillation columns, reactors, and more. "This work required close communication with engineers in the plant, as well as offsite contractors who were manufacturing the equipment," he says.

As operations supervisor of the Texas Marine Terminal, Shelton recalls that he found that his basic chemical engineering process and design knowledge "was instrumental in addressing problems associated with the transfer of chemicals from tankers, rail cars, barges, tank trucks, and pipelines to internal and external customers." In addition to daily operations, he also led several capital projects to upgrade terminal operations.

As he became more senior within the organization, Shelton also spent several years working at Union Carbide's global headquarters (first located in New York City and later in Danbury, Connecticut), rotating through various assignments. Each of these positions—including Business Analyst for the Silicones & Urethanes Intermediates Division, Manager of Financial Analysis for the Solvents and Intermediates Division, and Associate Director for Health, Safety & Environmental Protection Department with responsibility for Northeast based HS&EP Assessment Group—helped him to use his chemical engineering expertise and knowledge of the company's businesses to advance the company's overall corporate and business objectives.

REFLECTIONS ON BEING AN ENGINEER, AND A BLACK ENGINEER

"Throughout my career, I've found that teamwork is absolutely essential. As a process engineer, you're going to be responsible for operating a manufacturing process, which requires working with lots of folks—instrumentation people, pipe fitters, riggers, operating personnel, maintenance engineers, R&D people, lab people, and more," says Shelton. "You have to learn to work effectively with all types of people you are likely to encounter in the process plant… that's absolutely essential."

Shelton notes that there were few black students in his Chemical Engineering program at the University of Houston in the late 1960s, and he was one of the first black engineers to work at Union Carbide when he started in 1967. "Despite that, I never let my color be a barrier to me, and if it was an issue to others, I guess I was never very sensitive to it," he says.

"I always reached out and got to know the operators and maintenance and other technical folks and department heads and, because of that, I flourished," he says. "I think when you see the good in everybody they see the good in you."

"And importantly, when you demonstrate engineering competence, people trust you at the technical level and race becomes less of an issue—perhaps more so than in other career avenues," says Shelton. "I think that's unique to engineering—race seems to be much less of an issue, at least in my experience."

Throughout his career, Shelton had a long, satisfying relationship with the National Society of Black Engineers (NSBE). "It is gratifying to see the student-run organization, which had '7,000–8,000 members in 1986, grow to include more than 30,000 members today," he says. He served on the NSBE's National Advisory Board for more than 20 years; the Advisory Board advises the students who manage and run the organization.

"It was a good experience and I really enjoyed it—not just advising student leaders on the various projects they were planning, but also hearing about the challenges black students and alumni face to keep the membership pipeline strong," he says. "We developed better financial procedures and policies, developed additional NSBE chapters, developed long-range plans, and increased focus on academic excellence and NSBE's mission. The NSBE student leaders have a lot of energy and we made a lot of progress—we've come a long way and it's really been a labor of love for all of us."

REFLECTING ON HIS FRUITFUL CAREER

"I always had high hopes for my chemical engineering career," says Shelton. "My profession provided opportunities that exceeded my earlier expectations as a young, chemical engineering student."

He offers this advice for students: "chemical engineering is one of the most difficult of the engineering disciplines, and it should not come as a surprise that you will need to work very hard to achieve success in all aspects of the demanding chemical engineering curriculum," he says. "When times get tough, don't give up. I found that periodically studying with a small group of chemical engineering students gave me different insights on assigned problems."

Similarly, he notes that "it is important that you learn how to work in a team environment. In the plant, you will need to work closely and effectively with operators and maintenance personnel to achieve your goals. And it's very important to acquire a strong foundation in mathematics and chemistry in high school. This is a 'must' for success in chemical engineering."

"I am optimistic that the chemical engineering profession will continue to play a crucial role in adding value to improving our communities and lifestyles," says Shelton. "Chemical engineers will need to continue to make ongoing improvements in plant and process safety, and improve sustainability by helping to address global challenges ahead related to food production, water scarcity, and in creating sustainable sources of energy."

He also firmly believes that many emerging technologies—including bioengineering, smart manufacturing, process intensification, and more—will require the application of fundamental chemical engineering principles for success. "As such, our profession, in my opinion, is well positioned to make a positive impact on our society in the future."

Shelton strongly believes that "each of us should give something back—by robust participation in the AIChE, NSBE, and other professional organizations and mentoring younger engineers," he says, "to ensure our profession continues to provide value-added opportunities for future chemical engineers." Shelton co-chairs the AIChE Foundation's Minority Affairs Committee (MAC) Endowment Scholarship Fund and supports NSBE's fundraising initiatives to achieve its mission.

Looking back on his school experience, Shelton says: "My engineering professors were very effective in teaching me 'how to think' and solve problems through four challenging years of chemical engineering curricula. I was told that my engineering degree would hold the key to my profession and it did—it opened many doors. I have thoroughly enjoyed the technical diversity within our profession—there's really something for everyone." ∎

SEMICONDUCTOR INDUSTRY

In her own words
"Using chemical engineering to advance
the frontiers in semiconductor performance"

SUSAN WEIHER, PhD, GLOBAL FOUNDRIES

Current position:
Senior Director, Advanced Manufacturing Engineering Fab 1
Global Foundries (Bannewitz, Germany)

Education:
- BS in Chemical Engineering, University of California (San Diego)
- MS and PhD in Chemical Engineering, Stanford University

A few career highlights:
- Holds 15 patents, mainly in tungsten chemical vapor deposition (WCVD) and tungsten silicide chemical vapor deposition (WSiX)
- Instrumental in the introduction of the "Producer Platform" enabling technology for customers using CVD films in the 1990s (to revolutionize cost management by creating the now-famous "twin-chamber approach" that allowed the use of single gas lines for two chambers, thereby reducing costs during silicon wafer production from 30–50 wafers/hour to 60–100 wafers/hour that is now typical)
- Contributed pivotal ground work to the development of fully depleted silicon on insulator (FD-SOI) technology, that aims to revolutionize battery-powered consumer devices by creating ultra-low power requirements and extending battery life

"Throughout my school work and career efforts, my core focus has always been related to the engineering aspects of semiconductors," says Susan Weiher. For the past several years, she has worked at Global Foundaries, the second-largest foundry for semiconductor chips in the world. Prior to that, she spent 22 years with Applied Materials, responsible primarily for the manufacturing of semiconductor equipment that is used in chip fabrication.

Weiher's father was a "very proud physicist," and she and both of her siblings went on to earn chemical engineering degrees. Her brother is a polymer chemist, her sister works in materials engineering, and Susan's work focuses on semiconductor processes conducted under vacuum. "It's one of the things I love about chemical engineering—you're never locked into doing just the 'classical aspects' of the major," she says. "Throughout all my years working in the semiconductor arena, I've developed expertise and experience in many areas—plasma physics, atomic-level deposition, laser-measurement tools, and more—but I've been able to move through a lot of different positions and levels of responsibility."

During her nearly two decades with Applied Materials, she started out as a tungsten CVD engineer, and eventually moved into wafer inspection using laser-light inspection tools. Later, she took on a role in technical sales, managing a team of subject-matter experts. "People often rotate through these companies faster than you can print ID badges," she jokes, "but I was fortunate to stay at my first company for 22 years, in part because the opportunities to do something different and interesting were constantly present." She adds: "I don't think if I'd studied something different I would have had so many interesting opportunities."

MAKING THE RIGHT CHANGE AT THE RIGHT TIME

Weiher eventually moved to her current position about four years ago, with a desire to move from technical sales support into chip-manufacturing processes. The breadth and depth of her early career experiences "put me in a tremendous position to jump with enough base understanding to make a real contribution at my new company," she says.

Having worked for many years to support the technical sales team at her former employer, Weiher says "I understand what the tools actually do in the field and the way they function, and I really understand the needs and challenges of the end users of the equipment components that rely on semiconductor chips. With greater understanding of what the tools are actually doing, I am able to bring greater value to what the chips can—and should—be doing when they are used in products in the marketplace."

ON BEING A CHEMICAL ENGINEER

Weiher notes that the chemical engineering space marries so many interesting disciplines—chemistry, materials science, physics, math, and more—with engineering. Chemical engineers build a highly developed set of skills for complex problem-solving—developing an inherent way to understand and work on problems and to imagine many ingenious possible solutions. And, importantly, "they're able to really ponder the question and determine what else they need to figure things out," she says.

"As a chemical engineer, you study both physical and organic chemistry—so you really learn how molecules interact with one another," she says. "This has been very helpful in understanding semiconductor processes" throughout her career.

Such knowledge helps you to "develop a keen understanding of the way things behave, and get a clear picture in your head to really understand what's going on. You're able to develop a great instinctive understanding for how things respond in different spaces." And she notes that such problem-solving skills have broad applicability—not just in solving broad scientific and engineering challenges, but in managing large teams and complex projects. "I'm not sure if other disciplines provide the same set of problem-solving skills, and I've always had a blast using mine," she says.

With regard to her long career in semiconductor development, Weiher recalls the days back in the 1990s "when the semiconductor chip makers were the ones with the prestige and the money." Today, she notes, so much of the attention goes to the end products on the application side—smart phones and smart cars and the endless array of industrial and consumer products that rely on the semiconductor chips embedded within them.

"The chips themselves are now sort of taken for granted—seen as something of a commodity—but the truth is, the demands on semiconductor chips needed to meet future demand by consumers of them will force these chips to evolve in unprecedented ways. It's no longer just about making chips smaller and cheaper—our demand for more and more innovative products requires a revolution that must be driven by real genius," says Weiher. "Chemical engineers are uniquely trained to keep driving this frontier forward."

STAYING ON TRACK DURING COLLEGE AND GRADUATE SCHOOL

It's widely recognized that the chemical engineering curriculum involves a lot of hard classes, but the payoff is that it is also one of the most flexible and rewarding majors. Because of the challenges, many students will experience ups and downs while pursuing their undergraduate or graduate degrees. "Peer pressure and self-confidence are big issues for all students," she says. "You've got to learn the tricks to have to study better, and you should surround yourself with focused, hard-working peers," she says.

She says "pick a buddy at school who's smarter than you, become their best friend, and find ways to study together. Glue yourself to them, buy them coffee and doughnuts," she jokes. "Without this approach, I'm not sure I would have made it through school as successfully."

Similarly, Weiher says you should not let yourself become paralyzed by a fear of failure. "You're destined to fail at something along the way, and if you fail at something, well then you fail—it's all part of life's lessons. The goal is to use that experience to help you to refocus or redirect your efforts. Don't let a fear of failure stop you from taking on the full experience and trying new things at school or in the workplace."

ON BEING A FEMALE CHEMICAL ENGINEER

Weiher says she became interested in chemical engineering when her older brother was pursuing his chemical engineering degree. "Lots of people said 'don't take those classes, they're too hard,' but that must have motivated me more," she recalls. Eventually, both she and her younger sister earned degrees in chemical engineering.

"For reasons unknown, many young women are raised with better social skills (to navigate social situations) but many young women also feel more peer pressure about fitting in, and they may be more susceptible to stereotypical barriers," says Weiher. "I have no explanation why, but I didn't care much about what others thought as I was going through my chemical engineering schooling, or in my career." She continues: "My advice to young women engineering students is to do what's best to nourish your own intellect and curiosity and career objectives, and don't let the complainers or the naysayers get in your way."

"The folks who whine the most about how hard the courses are, well, they're the ones you should not be listening to," she adds. "If the group around you is not encouraging you to learn, you'll have a harder time learning."

The takeaway for young chemical engineering students, says Weiher, is this: "If you don't have curiosity in your studies and your career, you won't be successful. As a young person, you may not understand why you like something, but if you're curious about something, you'll find your way. Your education is not the end of the story—it is just the first chapter. When you combine curiosity with your education and evolving work circumstances, there's no end to what you can accomplish." ■

CHEMICALS SECTOR/PERFORMANCE ADDITIVES

In his own words
**"Applying engineering principles
to ensure safety and reduce risk"**

FREDERICK GREGORY, THE LUBRIZOL CORP.

Current position:
Process Safety & Risk Manager, Technical Fellow, The Lubrizol Corp. (Deer Park, TX)

Education:
- BS in Chemical Engineering, North Carolina State University (Raleigh, NC)

A few career highlights:
- Served as the technical lead at the company's new facility in Zhuhai, China, from initial conceptualization through construction and startup
- Named Technical Fellow at Lubrizol Corp.

Fred Gregory is Process Safety & Risk Manager, and a Technical Fellow, for Lubrizol, a global company that has two main divisions. Lubrizol's Additives and Advanced Materials Division manufactures performance additives for motor oils used in automotive, diesel, and marine applications, transmissions fluids, lubricants, and fluids, as well as coatings, inks, polyvinylidene chloride resins, and other advanced materials. Its Biological and Life Sciences Division produces tubing and plastic materials for medical applications.

Prior to joining Lubrizol 16 years ago, Gregory worked for two other chemical manufacturers (Hercules Co. for five years, and Ethyl Corp. for 12 years). "I have spent most of my career working as a process engineer, supporting plant operations, dealing with technical problems, retooling systems to accommodate different raw materials or produce new products, and managing the safety aspects of those process changes," he says.

"I also had summer positions during college with Morton Thiokol and Getty, both in support of operating plants, and even though I worked with very different types of chemistries, I gained skills and familiarity with similar types of chemical process equipment (agitators, filters, centrifuges, and more)," he adds.

During his tenure with Lubrizol, Gregory has also worked in a number of different chemistries, at different sites in different states and countries, and he moved into his current position as Process Safety & Risk Manager in 2013. "In this role, I don't really 'own' any of the projects any more—rather I look at all projects from the point of view of 'what could go wrong?' and try to build in extra safety features involving such things as supervisory control systems, sensors, interlocks, instrumentation, modeling and analytical tools, rigorous procedures, and more."

"We have a lot of younger engineers and employees. While they are certainly bright and capable, some do not have a lot of experience, so part of my role is to make sure that everything we are doing is being done as safely as possible," says Gregory. "With my experience in process engineering, I'm very comfortable working with process engineers in a mentoring role, and having a lot of 'what if' conversations."

PROCESS SAFETY MOVES TO CENTER STAGE

Gregory is happy to have witnessed an increased focus on process safety and risk reduction, and increased rigor and management commitment to safety, throughout the industry, in recent decades. "The industry as a whole definitely has less tolerance for incidents and is deploying a broader array of engineering tools and procedures to improving process safety," he says. "We are all expected (and willing) to shut down an entire unit until any safety concern can be properly addressed."

Today, through the use of greater automation and more sophisticated instrumentation, "we can see when valves were open and closed, and we have access to continuous pressure, temperature and level measurements," says Gregory. "The problem is almost having too much data and trying to decide how to get the most value out of the data so that we're able to get the most meaningful insight."

Reflecting on his role as Lubrizol's Process Safety & Risk Manager, Gregory says: "I'd worked as a technical manager at my earlier companies and had supervised people in those roles, but I think the hardest part for me in a managerial or supervisory role is having other people run a project because letting go is hard to do," he adds.

BROADENING HIS HORIZONS, LITERALLY AND FIGURATIVELY

Gregory is enormously proud of a challenging assignment he had a few years back that changed his perspective and his world view. Gregory spent five years working on the China Team, participating in the preliminary plant design, the detailed plant design, construction, and start-up of a new facility in Zhuhai, China (near Hong Kong and Macau), and spent three years in China during the project.

"I was the lead technical expert for two of the detergent units Lubrizol built at that facility," says Gregory. "It was a huge challenge and a great deal of fun to go from the proverbial paper design to starting up a major unit and producing product in another country—an undertaking most engineers are never fortunate enough to experience."

CONSIDERING THE BIG PICTURE

"My father had a strong technical background, having studied nuclear engineering and physics, and I was always good at math and science in high school," says Gregory. "I considered doing a double major in chemical engineering and mechanical engineering, because both seemed interesting to me, but most career options would have needed one or the other but not both, so I decided to focus on chemical engineering."

"Many young engineers feel like 'I have this job right now, so I'll have this job forever,' but that's just not true," says Gregory. "I tell my younger engineering colleagues that opportunities will come and go, and they should not be afraid to try something different—operations, R&D, marketing, sales, supply chain work. They all require chemical engineering expertise, and these broader opportunities will help you to ultimately decide where you want to focus, given your own skills and interests."

And given the natural tug between the "technical track" and the "managerial track," Gregory says, "good engineers don't always make good managers, and the opposite is also true. You have to let your network and your bosses know what you're interested in doing or trying; they may know about opportunities before you do and they will think of you when they do hear of something."

What's most important, he says, is to "always do your best job in your current position, so you can attract the right kind of attention and leverage to open other doors. No one wants to promote someone who isn't doing a good job," says Gregory. "The China opportunity came as a total surprise," he adds. "It has been one of my most important contributions and it completely changed my life." ∎

REFERENCES

Allen, David T. and David R. Shonnard, 2002. *Green Engineering – Environmentally Conscious Design of Chemical Processes*. Upper Saddle River, NJ: Prentice Hall PTR, Prentice-Hall, Inc.

Center for Chemical Process Safety. 2007a. *Guidelines for Risk Based Process Safety*. Hoboken, NJ: American Institute of Chemical Engineers, and John Wiley & Sons, Inc.

Center for Chemical Process Safety. 2007b. *Inherently Safer Chemical Processes: A Life Cycle Approach*. Hoboken, NJ: American Institute of Chemical Engineers, and John Wiley & Sons, Inc.

Sanders, Roy. 2014. *Chemical Process Safety: Learning from Case Histories*, 4th Edition. Woburn, MA: Butterworth-Heinemann.

Roles Chemical and Biomolecular Engineers Play

9

ChBE* ROLES IN THE LIFECYCLE OF A PROCESS PLANT

It is essential to consider all phases in the lifecycle of a process plant during the creation of the facility. (As discussed elsewhere in this book, the lifecycle of a chemical process plant starts with product and process selection, continues through the many steps required for design, construction, commissioning, and start-up; includes operation and maintenance; eventually includes shutdown and decommissioning of the facility; and, finally, includes restoration of the site.)

Chemical and biomolecular engineers play many essential roles in the lifecycle of a process plant (Figure 9.1). Some of these roles have been discussed in the previous chapters.

Process plants are at the core of the process industries, and chemical and biomolecular engineers provide the knowledge and expertise that is at the heart of all of these industries.

CHOICE OF DESIRED PRODUCT(S)

The selection of the desired product(s) and process(es) must be done carefully because it takes several years of work and lots of money to create a new production plant, or to modify an existing plant. Any product must be sold at a reasonable price and in sufficient quantity to make the investment profitable. It is usually assumed that a plant will continue manufacturing the product for 20–30 years or longer.

In addition, the choice of product and manufacturing process are critical to making the product safely, with zero or minimum environmental impact. One or more chemical or biomolecular engineers are typically key team members because of their knowledge of the properties of materials and of chemical or biochemical manufacturing processes. Typical activities include:

- Market research on existing and potential markets, including competitive products
- Screening of potential products for desirable properties and low cost
- Choice of products that will have zero or minimum safety or environmental risk in use and manufacturing

PRODUCT RESEARCH, DEVELOPMENT, AND ENGINEERING

After a preliminary choice of product type has been made, chemical and biomolecular engineers may work with chemists, biochemists, and other project team members on research, development, and engineering of the product(s). Research goals may include optimization of the physical and chemical form of the product, selection of the best raw materials for production of the product, and prediction of the fate of the product in use and in the environment (Figure 9.1).

As mentioned in Chapter 8, **green engineering** starts with efforts to select the product (and any chemicals that are used, produced as byproducts or waste streams) for favorable properties from an environmental aspect (Figure 9.2).

---

* Chemical and Biomolecular Engineering.

FIGURE 9.1   Chemical engineering efforts to produce new products and the new processes that may be needed to produce them typically begin with fundamental laboratory research. (Source: Shutterstock ID 730300147)

FIGURE 9.2   Shown here is an environmental engineer inspecting a large clarifier that is used to treat and purify industrial wastewater by removing suspended solids by sedimentation so that the treated wastewater can either be safely discharged to the environment (per the facility's environmental permit regulations) or, increasingly, recycled back to the inlet of the facility, so it may be used again and again. (Source: Shutterstock ID 153095864)

An **Inherently safer design** (discussed in Chapter 8) starts with selection of the product based on its safety and on the safety of the process and materials used to produce it.

PROCESS RESEARCH AND DEVELOPMENT

Process research and development is focused on the safe production of the product. If the decision has been made to use an existing process technology to produce the product, then most of the process research has already been done. If not, then the chemical research engineer may conceptualize several different, competing processes to produce the product and then conduct research at laboratory- or pilot-plant-scale to gather test data on the proposed process(es).

FIGURE 9.3   Biomolecular engineers with deep knowledge of biochemical and biomolecular engineering principles and objectives play a key role in developing and reviewing engineering drawings for new pharmaceutical plants. (Source: Shutterstock ID 582999514)

PRELIMINARY PROCESS SELECTION, CONCEPTUAL ENGINEERING, DESIGN, AND COST ESTIMATE

Based on known technology and laboratory data the *chemical process engineer* will develop preliminary designs for one or more processes to produce the desired product. He/she will probably use *process simulation* and *process synthesis* (discussed in Chapter 8) to optimize the design of one or more processes, in terms of one or more target objectives, such as maximizing yield of the process, minimizing unit costs, minimizing unwanted byproducts or regulated effluents, and so on.

The sequence of unit operations and reaction steps in a process is typically documented with a series of drawings or computer models, each one called a *process flow diagram* (PFD). The PFD also shows the calculated temperatures, pressures, and compositions of the various process streams that connect the steps in the process (Figure 9.3).

Preliminary estimates will be made of the type, size, and cost of the various process equipment, vessels, piping, process control elements, and so on in the process, to enable the best design to be obtained. Throughout this process, the *process safety engineer* (usually a chemical engineer) will ensure that proven and widely accepted *process safety management* (PSM) practices are applied. PSM practices aim to identify potential process hazards and to minimize them by working with the chemical process engineer and other members of the project team.

The *chemical process engineer*, and the *process safety engineer* will jointly apply the principles of inherently safer design, with active participation of the other members of the project team. The process safety engineer may conduct computer simulations to evaluate the likelihood and impact of potential hazardous incidents that could occur during the operation of the facility, such as fires, explosions, or releases of toxic gases. Such modeled insight gives the design team relevant information that can then be used to systematically develop prevention and mitigation measures. These measures involve both process engineering, the use of appropriate safety systems, and the proper operational and managerial practices with regard to operation and ongoing maintenance efforts, as well as appropriate environmental health and safety practices.

The chemical process engineers and the *environmental engineer* (often a chemical engineer) will work together with other members of the project team to select and design an environmentally safe and clean process.

SITE SELECTION

If a site has not already been chosen at this point, the project team will evaluate potential sites for their suitability for the process. The chemical process engineer, the environmental engineer, and the process safety engineer will work with other members of the team to consider factors required by the process, such as available land area, availability of raw materials and qualified personnel, access to transportation and utilities (such as sources of water and power/electricity), cost, permitting issues, proximity to the public, and access to public facilities like schools and hospitals. For a billion-dollar, world-scale facility, hundreds of acres of land may be required, and the team of technical and business decision makers will be quite large. Also, it there are process hazards that could potentially have offsite consequences, it may be wise to acquire additional land located around the plant site (to provide a generous buffer between the facility and the surrounding community). Such land acquisition also allows for future plant expansion.

PERMITTING

The environmental engineer and the chemical process engineer will support the corporate legal and regulatory group in discussions with local, state, and national regulatory agencies and with other stakeholders to obtain permits to build and operate the plant. Efforts to secure regulatory permits need to start early in the project, because permits often take months or even years to complete, and delays in completion of the project are often costly (delaying the onset of production and forestalling the time when the product can finally enter the market to start generating sales income for the company).

The chemical process engineer and the environmental engineer will work together to predict the flowrates and compositions of all potential emissions from the plant, so they can be incorporated into the federal, state, and local environmental permit applications. Chemical and mechanical engineers from the companies that will be supplying the engineered equipment may also provide designs of specialized equipment and use modeling and simulation tools to predict emissions after suitable environmental controls have been applied. The environmental engineer may also use computer models to predict the atmospheric dispersion of any emissions to ensure environmental safety.

In some cases, a specialized environmental consulting firm with chemical and environmental engineers may be hired to work with the project team to conduct the environmental modeling and permitting.

FRONT-END ENGINEERING DESIGN AND PLOT PLAN DEVELOPMENT

A *plot plan* is a drawing or computer model that shows the physical location and arrangement of the various vessels, equipment, and buildings in the planned process plant. The process engineer usually develops the plot plan in conjunction with the piping designer, who creates the computer model of the plant layout.

Equipment arrangement is very important. For example, the process piping and instrument cables that connect the various parts of the process may cost more than the vessels and equipment used in the process. Therefore, the arrangement of process equipment can have a dramatic effect on piping costs, and careful planning of equipment spacing and layout can help to reduce some of the overall project costs. In most cases, the various parts of a process should be in the same order as their order in the process. In addition, process safety considerations have a very important impact on plant layout (Figure 9.3).

The control room and the administrative building should be located at safe distances from any high-pressure and/or highly reactive chemical processes. But the control room must also be located to give good access to the process for the plant operating personnel. Also, if there is a possible loss of containment of a toxic gas, that part of the process should be located on the downwind side of the plant, and the control building should have a protective air conditioning system.

Plot plan reviews must be conducted by a project team with expertise in all aspects of the plant, including operations and maintenance personnel.

Front-end engineering design (FEED) is done in parallel with plot plan development and is led by the *project manager* (also often a chemical engineer) and the *lead process engineer*, supported by other chemical process engineers and by other members of the project team.

Chapter 8 outlines the steps between preliminary engineering and plant start-up. Chemical process engineers play important roles in each phase. Some chemical engineers specialize in detailed process engineering, while others may specialize in start-up of many different engineering aspects of the overall plant function, operation, and support systems (such as control systems, environmental and waste-treatment systems, and more). Start-up teams also almost always include at least one chemical process engineer from the project design team.

OPERATION, OPTIMIZATION, AND MAINTENANCE

The new plant operations and maintenance teams take over the running of the plant during and after start-up. These team members may continue working at that plant for most of their careers, taking on roles of increasing responsibility as they gain experience and expertise, or they may transfer to leadership roles at other plants.

Chemical engineering roles in operations include supervising plant operations, diagnosing problems in the operating plant, carrying out computer modeling, designing and operating process control systems, undertaking efforts to optimize the plant, selecting improved technology and equipment, and guiding plant expansions. Chemical engineers with experience in each manufacturing process in an existing plant may also be called upon to help apply that process technology at a new plant, by temporary membership on the project design team and/or on the startup team.

Maintenance team members are often mechanical engineers, mechanics, instrumentation engineers, and other technicians (Figure 9.4).

CLOSURE, RECYCLING OF EQUIPMENT, AND RESTORATION OF SITE

Technology continues to advance and, at some point, many older plants eventually become obsolete due to ongoing advances in new and improved products, processes, process equipment, and instrumentation. Chemical engineers play an important role in the evaluation of existing plants and in the pursuit of potential opportunities to modernize them. However, if modernization of an existing chemical manufacturing unit is not practical, then chemical engineers participate in the final shutdown, cleaning, and mothballing or sale of any still-useful vessels and equipment. Process vessels, piping, and other materials that cannot be reused may be sold for scrap.

The chemical process plant site may require remediation due to contamination by any process materials leaked or spilled during plant operations. Chemical/environmental engineers often lead the needed environmental remediation work, which typically requires removal and landfill of contaminated soil and the application of various process technologies to purify contaminated groundwater.

INDUSTRY VOICES

"As a trained engineer, I have a whole other lens through which to see the world, and I have garnered respect from my peers, professors and colleagues from the time I enrolled in law school with an engineering degree, and what I learned still applies to the many of the cases I work on today."

—Anita E. Osborne-Lee, Attorney-at-Law (Private Practice), page 159

FIGURE 9.4   Shown here is the installation of the channel head of a shell-and-tube heat exchanger by engineering and maintenance personnel. (Source: Shutterstock ID 726456217)

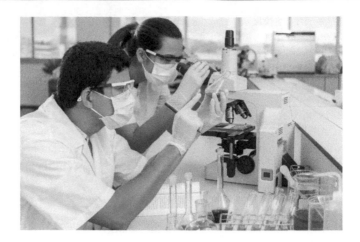

FIGURE 9.5   Biomolecular engineers are shown here at work in a laboratory in pursuit of ongoing advances in pharmaceutical and bio-based chemicals developments. (Source: Shutterstock ID 451124581)

ChBE ROLES IN BIOLOGICAL ENGINEERING

Synthetic biology can be broadly defined as the use of technologies that accelerate the processes of genetically engineering biological systems (Levesque-Tremblay and Schuster 2017).

Synthetic biology is just one area of biological engineering—but a very important one. Because biological engineering is in its infancy, the roles of the biological engineer will likely have a major research-and-development component in the years to come (Figure 9.5).

The career of Marc-Andre D'Aoust is an excellent example. In 1999, he was the seventh employee of Medicago, a biotechnology start-up. Now Medicago has 300 employees and D'Aoust is co-inventor on 330 patents or patent applications (*Chemical Engineering Progress* 2017). The pioneering vaccine technology being developed by Medicago has impressive possibilities. In brief, the genetic sequence of a pathogenic virus is used to create a gene construct that defines a virus-like particle (VLP). The VLP has none of the genetic material of the virus that enables it to replicate in a host cell, but it does have the genetic code of the virus, which creates the non-infectious outer antigenic proteins and other biomolecules that stimulate the immune response of the person receiving the VLP in a vaccine.

The production of a new vaccine based on this innovative approach is very inventive. The gene construct of the VLP is grown in an *Agrobacterium* and then transferred into a plant (*Nicotiana benthaminana*) that is grown in an automated system to produce commercial quantities of the VLP. The VLPs are harvested from the plant, and then purified to regulated vaccine quality.

The engineering steps required to carry out this biological process, which involves managing the virus, the bacterium, the plant, and the purification system, are quite complex. This process involves a variety of novel unit operations, which can be a challenge, even for inventive biological engineers.

At the time of writing, the H5N1 pandemic influenza VLP vaccine from Medicago has now completed phase II human clinical trials. Other products under development by Medicago include seasonal influenza and H1N1 vaccines, a non-influenza VLP, and a rabies vaccine. Shih (2017) has written an excellent history and overview of the use of specialized living plants to produce bioproducts.

ChBE ROLES IN BIOMOLECULAR ENGINEERING

Biomolecular engineering was defined in Chapter 1. For convenience, that definition is repeated here.

In addition to using many of the chemical engineering unit operations and separation processes, *biomolecular engineering* also employs biological and biochemical sciences to develop processes to produce biomolecular products (such as medicines, fuels, amino acids, vitamins, and others) using unique chemical and biomolecular engineering principles. These bioproducts often involve biochemical transformations using enzymes, living cells, or plants in the manufacturing process.

Biomolecular engineering is a subset of biological engineering, with the emphasis on the manufacture of biomolecular products. There has already been substantial progress in modifying bacteria, yeast, and mammalian cells by inserting genes that encode the synthesis of specific bioproducts of interest.

One example is a process that involves the insertion of a gene to produce insulin (used to help patients with diabetes to manage their blood-glucose levels) inside a species of bacteria. The genetically modified bacteria are then grown inside sterile bioreactors to produce insulin. After the fermentation reaction is complete, the bioreactor contents are processed to recover and purify the insulin. This is superior to the historic process, which relies on harvesting insulin from animals.

Like biological engineers, the role of the biomolecular engineers often has a large research-and-development component. However, the engineering of cellular bioreactors for cell culture is much further advanced than the engineering of VLP manufacturing systems.

ChBE ROLES IN BIOMEDICAL ENGINEERING AND MEDICINE

Biomedical engineers are employed in several industry sectors, all related to health (Madhavan, Oakley, and Kun 2008; Linsenmeier and Gatchell 2008):

- Invention and development of implantable and extracorporeal (outside the body) medical devices (e.g., pacemakers, dialyzers, intraocular lenses)
- Development of hospital diagnostic systems and other medical products (e.g., echocardiogram (EKG) heart monitors, automated blood analyzers, catheters, surgical tools)
- Development and manufacturing of medical imaging technologies (e.g., optical, magnetic resonance, computed tomography, and more)
- Design and manufacture of prosthetic devices, orthotic devices, and other rehabilitation aids
- Development, demonstration, and manufacture of novel pharmaceuticals

There is some overlap between the undergraduate curriculum of biomedical engineers and ChBEs. Of the 17 key concepts in accredited biomedical engineering programs in the United States (Madhavan, Oakley, and Kun 2008; Linsenmeier and Gatchell 2008), 14 would also be in a typical chemical engineering curriculum. Chemical engineers transitioning into biomedical engineering would need to take several fundamental courses in the life sciences and in the mechanical properties of biological tissues. Biomolecular engineers, having already had several life sciences courses, would need to learn about the mechanical properties of biological tissues, as well as human physiology.

Those seeking to become biomedical engineers may wish to earn a master's degree (MS) or a doctorate degree (PhD) in the field to get additional course work and to facilitate their work in research and development.

Either chemical or biomolecular engineering would be adequate preparation for entry into a medical school; however, again, additional knowledge of life sciences would also be required.

ChBE ROLES IN MIDDLE AND CORPORATE MANAGEMENT

Engineering management roles are mentioned elsewhere in this book. Examples of supervisory or middle management roles include the following:

- Leading a team of chemical process engineers in the design of a billion dollar-scale process plant by an outside (third-party) engineering, procurement, and construction (EPC) contractor
- Serving as project engineer or project manager for a major EPC project (see Figure 9.6)
- Managing all the chemical process engineers in an EPC contractor

FIGURE 9.6 Project management encompasses a variety of activities that must be faithfully executed if a project is to succeed. (Source: Shutterstock ID 36480153)

- Managing the entire engineering section in an EPC contractor
- Serving as group leader for a small research and development team
- Serving as production manager of a manufacturing unit in a process plant
- Serving as plant manager
- Leading the plant engineering function in a process plant

These first-line supervisor and middle-management roles need job knowledge in chemical or biomolecular engineering.

Corporate and other upper-management positions typically have a larger scope. The following are examples of upper-level corporate management positions that benefit from an individual with broad and deep ChBE knowledge:

- Plant manager for a large process plant
- Vice president for production, responsible for all process plants in the company
- Vice president for research and development
- Vice president for health, safety, and environment
- Vice president for projects in an EPC contractor

Occasionally, chemical and biomolecular engineers take on corporate management positions where their ChBE background is helpful (in terms of truly understanding the scope of the business, the potential market opportunities, and technical or market challenges that exist), but is not necessarily required:

- President of the company
- Chief executive officer (CEO)
- Executive vice president
- Chief operations officer (COO)
- Chief financial officer (CFO)

Many chemical engineers who seek to go the management route choose to earn a master of business administration (MBA), or a degree in accounting or finance to supplement their ChBE job knowledge. This would provide training in accounting, business, finance, and management. With such a combination of degrees (and the knowledge and skills attendant in both), the graduate could then focus his or her career more precisely and present themselves as an even-more-valuable candidate for a desired position within the organization.

ChBE ROLES IN ENERGY ENGINEERING

Energy engineering or *energy systems engineering* includes the following areas:

- Energy efficiency
- Energy services
- Facility management
- Plant engineering
- Environmental compliance
- Alternate energy technologies

Chemical engineering has enough overlap with these areas that one could transition into a utility company (such as electric power or natural gas distribution), or an energy management role in a process plant.

Petroleum engineering is also related and will be briefly described below.

ChBE ROLES IN ENVIRONMENTAL ENGINEERING

Many chemical engineers choose to specialize in environmental engineering. With such added specialization, they can work with business, government, and industry. Their core role is to work with chemical process engineers first to identify, characterize, and quantify the various emissions, effluents, hazardous waste, and solid wastes generated by a process plant. Next, they work with chemical process engineers to modify and optimize the design to eliminate or minimize emissions, effluents, hazardous wastes, and solid wastes generated by the process. Then they work with chemical process engineers and equipment vendors to design and specify the various types of process environmental equipment needed to ensure that anything leaving a plant will have a minimal effect on the environment (Rubin and Davidson 2001). Such equipment components include the following:

- Absorbers/scrubbers to remove acid gases from power plant emissions
- Electrostatic precipitators and fabric filters to remove particulates from power plant emissions
- Sedimentation and clarifier systems to remove suspended solids from wastewater streams
- Chemical precipitation and biological wastewater-treatment systems to remove dissolved solids from wastewater streams
- Creation of a manifest system for tracking the origin, treatment, transportation, and ultimate disposal of all hazardous wastes

Environmental engineers routinely work with the legal department, with internal environmental engineering experts at the company, and often with a third-party (external) environmental consulting firm, to prepare complex state, local, and federal permit applications, and they provide advisory services and input in meetings with regulatory authorities. They may also oversee operation and maintenance of the environmental equipment that is operated by plant production and maintenance personnel.

Other environmentally oriented chemical engineers may work in environmental consultancies, regulatory agencies, and in EPC (engineering, procurement, and construction) contractors.

ChBE ROLES IN FOOD ENGINEERING

Food engineering is a multidisciplinary field that combines microbiology, applied physical sciences, chemistry, and engineering for the food, beverage, animal feed, and related industries.

Chemical engineers can easily seek careers with a focus on food engineering with some additional coursework and training in microbiology, biochemistry, and food technology. Biomolecular engineers can also transition to food engineering roles relatively easily, needing only training in food technology.

Engineering for the design and construction of a food processing plant shares many similarities with engineering efforts that are needed for a traditional chemical plant, but with the addition of the types of process operations that are needed to produce high-quality, sterile foods and beverages and provide adequate packaging for them. The types of added process operations include, for instance, pasteurization or sterilization of the processed materials, and aseptic processing equipment to prevent microbial contamination and spoilage. Refrigerated processing is also often employed to prevent thermal degradation, and a wide array of packaging options is available to allow the foods to be protected and transported over long distances and stored for long periods of time.

ChBE ROLES IN GOVERNMENT

Chemical and biomolecular engineers play many important roles in government.
Examples of federal agencies include:

- Basic research (National Science Foundation (NSF), National Institutes of Health (NIH), National Bureau of Standards (NBS), National Laboratories (NL))
- Energy research (Department of Energy (DOE), NL, NSF)
- Environmental research (Environmental Protection Agency (EPA), NSF, NIH, DOE)
- Medical and health research (NIH, EPA, Food and Drug Administration (FDA))
- Energy regulation (DOE, Federal Energy Regulatory Commission (FERC))
- Environmental regulation (EPA, various agencies through Environmental Impact Statements)
- Food regulation (FDA, Department of Agriculture (DOA))

There are numerous state and local agencies that play important regulatory roles.*

ChBE ROLES IN HEALTH, SAFETY, SECURITY, AND ENVIRONMENTAL MANAGEMENT

Most industrial organizations have several career opportunities related to the management of health, safety, security, and the environment (HSSE). Examples of such positions are given below.

- *HSSE manager for an EPC project.* Responsible for ensuring that the entire project team engages in safe and healthy lifestyles, that the process plant will be safe to build and operate, and that the plant will not have adverse effects on the environment; also responsible for ensuring that there is an adequate physical security and cybersecurity plan in effect.
- *Vice president for HSSE for an EPC contractor.* Responsible for ensuring that the HSSE manager on each project is meeting or exceeding all HSSE objectives; also responsible for meeting all other HSSE objectives of the company.
- *HSSE manager for a process plant.* Responsible for ensuring that the entire plant team (including both direct employees and contractors) engage in safe and healthy lifestyles, and that the process plant operates safely with no adverse effects on the environment; also responsible for ensuring that there is an adequate physical security and cybersecurity plan in effect to prevent and deter terrorist attacks.
- *Vice president for HSSE for a company.* Responsible for ensuring that all company operations are carried out safely with no adverse effects on the environment; also responsible for meeting all other HSSE objectives of company.

ChBEs can transition into HSSE roles by attaining proficiency in HSSE practices.

---

* One of this book's authors, Victor Edwards, worked at the US National Science Foundation in Washington, DC in 1971, on a leave of absence from the faculty of Cornell University. As Associate Director of the Engineering Chemistry Program and Manager of the Enzyme Technology Program at NSF, he helped start the first national research program in industrial biotechnology.

ChBE ROLES IN LAW

INTELLECTUAL LAW FOR NEW PROCESSES AND NEW PRODUCTS

The knowledge and experience of chemical and biomolecular engineers qualify them for evaluating the originality of new products and processes that are being proposed for patent protection under the law. The test for originality of an invention is that its creation would not have been apparent to a person skilled in the art; thus, the strongest patent attorneys are those who also have an underlying degree in the discipline for which they are routinely evaluating patent proposals (on behalf of the government patent office) or helping inventors to develop thorough patent applications. ChBEs can either work with a patent lawyer when applying for a new patent or trademark, or they can attend law school and prosecute patent applications themselves as a patent attorney.

ChBEs (with or without having attended law school) are also often called upon to provide expert legal testimony in their area of expertise in legal cases involving alleged patent violations and other intellectual property issues.

REGULATORY AFFAIRS AND PERMITTING IN THE PROCESS INDUSTRIES

The knowledge of chemical and biomolecular engineers also qualifies them for preparing process information required for permit applications and for meeting with regulatory officials. Again, it is usually preferable either to work in partnership with an attorney skilled in permitting, or to become qualified in the law. One of this book's co-authors (Vic Edwards) led the 1978 technical preparation and presentation of a successful application and hearing before the US Federal Energy Regulatory Commission for approval of the US$12.8-million Research, Development, and Demonstration Project to Convert Municipal Wastes and Biomass into Synthetic Natural Gas.

CORPORATE LAW, ESPECIALLY IN THE CHEMICAL PROCESS INDUSTRIES

A ChBE background, in combination with a law degree, provides a powerful combination and sound preparation for the practice of corporate law for a wide array of companies throughout the many segments of the chemical process industries.

If one decides to enter law school after gaining a few years of workplace experience within the ChBE in industry, that experience would be valuable for corporate law.

ChBE ROLES IN MATERIALS ENGINEERING AND METALLURGY

Materials engineering refers to the selection of the correct materials for the application for which the part or component will ultimately be used. This material selection process includes choosing the material, paying attention to its specific type or grade based on the required properties. Some important aspects include choosing a material that will: provide the needed mechanical strength and structural integrity; be able to withstand the anticipated range of temperatures and pressures to which the component will be exposed; and be able to withstand potential damage from corrosive attack by the fluids or environmental conditions the part may contact.

Materials science is the scientific discipline related to the structure/properties of all materials: metals, ceramics, polymers, and so on.
Most chemical engineering undergraduate programs include an introduction to materials engineering and/or material science. The transition from a chemical engineering role to a materials engineering role is not unusual.

Metallurgy is a domain of materials science and engineering that studies the physical and chemical behavior of metallic elements, their inter-metallic compounds, and their mixtures, which are called alloys.

ChBE ROLES IN PETROLEUM ENGINEERING

Petroleum Engineering is a field of engineering concerned with the discovery, recovery, and production of hydrocarbons (also called fossil fuels), which can be either oil and/or natural gas, that exist in complex underground geological reservoirs.

Once oil-and-gas reserves are discovered beneath the surface of the earth, petroleum engineers work with geoscientists and other specialists to understand the geologic formation of the rock containing the reservoir. They then determine the drilling methods, design the drilling equipment, implement the drilling plan, and monitor operations.

Petroleum exploration and production are deemed to fall within the "upstream" sector of the oil-and-gas industry; by contrast, "downstream" operations relate to the use of chemical engineering (and other) principles to convert the recovered petroleum and natural gas reserves into thousands of chemical, energy, plastics, and other intermediate and final products that are widely used throughout society. Many chemical engineers have successfully transitioned into petroleum engineering roles by acquiring knowledge of geology, of petroleum technology, and of drilling methods and equipment.

ChBE ROLES IN PROCESS SAFETY MANAGEMENT

Process safety management (PSM) is a management system that is focused on prevention of, preparation for, mitigation of, response to, and recovery from catastrophic releases of chemicals or energy from a process associated with a facility (Center for Chemical Process Safety 2007).

PSM was created in the mid-1980s in response to the catastrophic incident involving the atmospheric release of more than 40 tons of the toxic chemical methyl isocyanate (MIC) in Bhopal, India—a catastrophe that killed at least 4,000 people and injured more than 100,000 in 1984.

The development of the PSM movement has been led by the international chemical engineering community, working with operating companies throughout the global chemical process industries, and building on the foundation of loss-prevention efforts that were formally initiated in the 1960s.

The AIChE and the Center for Chemical Process Safety (CCPS), along with the Health and Safety Executive (HSE) and the Institution of Chemical Engineers in England, have been among the leaders in the development of this chemical engineering specialty. To date, CCPS has published more than 100 books on various aspects of process safety, and there are many other excellent books by other authors and organizations. Table 9.1 lists the 20 elements of risk-based process safety.

Chemical engineers involved in PSM often have the direct responsibility of ensuring process safety of an entire operating plant or an entire plant design project. Many PSM professionals either bring years of operating experience to the process industries, or they earn a graduate degree in chemical engineering, specializing in process safety.

ChBE ROLES IN PROJECT MANAGEMENT

A *Project manager* is responsible for the planning and execution of a project.

The following description of the ideal project manager is adapted from the website of the Project Management Institute (Project Management Institute 2017):

A project manager should:

- Be organized, passionate, and goal-oriented
- Understand what projects have in common
- Understand the strategic role that the project plays in how organizations succeed, learn, and change

TABLE 9.1   **Risk-Based Process Safety (RBPS) Elements**

Commit to Process Safety

Process safety culture
Compliance with standards
Process safety competency
Workforce involvement
Stakeholder outreach

Understand Hazards and Risk

Process knowledge management
Hazard identification and risk management

Manage Risk

Operating procedures
Safe work practices
Asset integrity and reliability
Contractor management
Training and performance assurance
Management of change
Operational readiness
Conduct of operations
Emergency management

Learn from Experience

Incident investigation
Measurement and metrics
Auditing
Management review and continuous improvement

Source: Center for Chemical Process Safety, Guidelines for Risk Based Process Safety, Hoboken, NJ: American Institute of Chemical Engineers and John Wiley & Sons, Inc., 2007.

Project managers are change agents. They envision and create project goals of their own and they use their skills and expertise to create a sense of shared purpose within the overall project team (Figure 9.6). They enjoy the new challenges and the responsibilities of driving business results while safeguarding plant personnel, the surrounding community and environment, and protecting plant assets.

Project managers work well under pressure and are comfortable with change and complexity in dynamic environments. They can shift easily between the "big picture" and the small-but-critical details, knowing when to concentrate on each.

Project managers cultivate the people skills needed to develop trust and communication among every project stakeholder; this includes its sponsors, those who will make use of the project's results, those who command the resources needed, and the project team members.

Over the course of their careers, project managers develop and acquire a broad and flexible toolkit of techniques, allowing them to resolve complex, interdependent activities into tasks and subtasks that are documented, monitored, and controlled. They adapt their approach to the context and constraints of each project, knowing that no "one size" can fit the great variety of projects. And they are always improving their own and their teams' skills through lesson-learned reviews at project completion. Project managers are found in every kind of organization—as employees, managers, contractors, and independent consultants. With experience, they become *program managers* (responsible for multiple related projects) or *portfolio managers* (responsible for selection, prioritization, and alignment of projects and programs with an organization's strategy), among others.

Project managers are in increasing demand worldwide. For decades, as the pace of economic and technological change has quickened, organizations have been directing increasing amounts of their energy into projects rather than routine ongoing operations (Project Management Institute 2017).

ChBE ROLES IN NANOTECHNOLOGY
AND MICROELECTRONIC ENGINEERING

Many would agree that the invention of the transistor in 1947 and the invention of the integrated circuit in 1958 have transformed society. Ever since, integrated circuits have steadily become both smaller and far more complex, enabling them to do more.

Microelectronic engineering involves the micro-scale and nanoprocessing- and nanotechnology-based techniques that are required during the fabrication of electronic, photonic, bioelectronic, electromechanic, and fluidic devices and systems, and their application in the broad areas of electronics, photonics, energy, life sciences, and the environment.
Microelectronics are typically electrical devices made of integrated circuits that are fabricated on a micro-scale, with dimensions of micrometers (Figure 9.7).

Microelectronics is the design, manufacture, and use of *microchips* and *microcircuits*.

A *microchip* is a tiny wafer of semiconducting material used to make an integrated circuit.

A *microcircuit* is a minute electric circuit, especially an integrated circuit.

The ever-shrinking size and power of microchips has led to computer technology with a scale of less than 20 nanometers. Widely proven developments in microelectronics have given way to pioneering breakthroughs that involve the manipulation of materials on a nano-scale. This is no longer microelectronics; it is nanotechnology.

Nanotechnology is the branch of technology that deals with dimensions and tolerances of less than 100 nanometers—within the realm of nanotechnology, one nanometer refers to one-billionth of one meter—especially the manipulation of individual atoms and molecules (Shelley 2005). Nanotechnology refers to the ability to synthesize, manipulate, and characterize matter at particle dimensions of 100 nm or less. To give a sense of the perspective of just how small this is, consider that one micrometer is equivalent to 1,000 nanometers, and the diameter of an average human hair is about 10,000 nanometers (Shelley 2005). When chemical engineers can produce and manipulate matter at these incredibly tiny dimensions, they are able to exploit a broad array of novel, size-dependent properties and phenomena. For instance, when produced at infinitesimally small particle sizes, materials as varied as metals, metal oxides, polymers, ceramics, carbon derivatives, and others have extraordinary ratios of surface area to particle size. Thus, optical, electrical, and magnetic properties, as well as hardness, toughness, and melting point, can differ markedly from the same materials at larger scales (Shelley 2005). Chemical engineers can lend their expertise to the ongoing development and exploitation of nanotechnology-related advances.

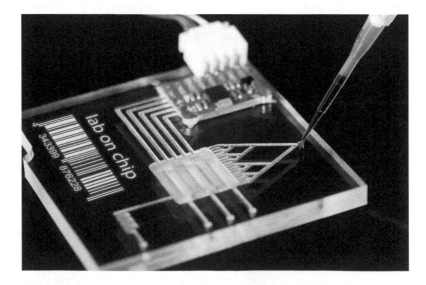

FIGURE 9.7   A "lab on chip" is a type of microelectronic device that integrates several laboratory processes in one device. (Source: Shutterstock ID 281615639)

The careers of Susan Weiher (profiled in Chapter 8) and Chris Bowles (profiled in Chapter 9) illustrate how chemical engineers can become great contributors to microtechnology and nano-technology products.

ChBE ROLES IN INDUSTRIAL RESEARCH AND DEVELOPMENT

Chemical and biomolecular engineers play a wide variety of roles in industrial research. While the list is endless, a few illustrative examples include the following:

- Research on the microbial production of ethanol from cellulosic biomass
- Research on new flexible organic solar cells
- Research on microbial insecticides that can kill many crop pests but remain harmless to beneficial plants, animals, and humans
- Research on new batteries and other energy-storage systems

Often, small teams may conduct research and development, with each team member providing a different specialty. For example, in addition to a biomolecular engineer on the team to create a new microbial insecticide, a bacteriologist and an entomologist (insect scientist) would add needed breadth and depth in the pursuit of breakthroughs related to the discovery of new species of bacteria that are pathogenic to crop pests but harmless to beneficial plants, humans, and animals.

Management roles in research and development include a *lead researcher* for a group of specialists working on a specific project, or a *research director* who directs all research projects in a company.

ChBE ROLES IN UNIVERSITY TEACHING AND RESEARCH

There are at least three roles that a ChBE professor typically plays at a research university:

- Teaching undergraduate and graduate students
- Conducting independent research, and in collaboration with their own graduate students, and with peers or colleagues at the institution, other institutions, or in industry
- Providing support and service to the profession and to the intellectual community related to the profession

The education of undergraduates is mostly based on an organized four-year curriculum of required courses in the topics listed in Chapters 4 through 8. ChBE professors usually teach ChBE undergraduate courses, although graduate students may lead problem-solving sessions.

Graduate students usually have a core of courses in the first year or two, but research is an important part of their education. Independent research is essential for doctoral candidates. The role of the ChBE professor is to serve as an expert guide for research ideas and analysis.

ChBE professors are expected to raise funds for research, usually from the government or from industry. They are also expected to be recognized for their contributions in their field of endeavor through their publications of important research results. Professors who spend their careers at one or more research universities typically take great pride in "helping to launch" legions of graduate students through the undergraduate and graduate-level courses they teach, the graduate-school work on which they collaborate, and ongoing personal and professional mentoring that often extends beyond the college and graduate-school years.

Another expectation of the ChBE professor is that they should provide service to the community, such as in the form of volunteering in professional societies like the AIChE and the American Chemical Society (ACS), or by consultation with industry and government in their area of expertise.

BIOPROCESSING/PHARMACEUTICAL DEVELOPMENT

In her own words
"Building a career in the biopharmacuetical industry
with a solid chemical engineering foundation"

SHEILA (YONGXIN) YANG, Genentech/Roche

Current position:
Principal Process Engineer, Pharma Technical Asset Americas, Genentech/Roche

Education:
- BS Chemical Engineering, Hunan University
- MS Chemical Engineering, University of British Columbia, Canada

A few career highlights:
- Served as Project Engineer/Process Area Lead for a number of mid- to large-sized capital projects
- Serves on the Editorial Board for *Chemical Processing* magazine

At Genentech, Sheila (Yongxin) Yang's title is Principal Engineer and her professional work is focused on the engineering, design, and construction of highly specialized facilities that produce biopharmaceuticals using biological processes. Whether they involve building a brand-new facility or modifying an existing one to produce new products or use new processes, her work typically relates to both the design of the facility itself, and the design, selection, purchase, installation and troubleshooting of the necessary equipment and associated systems and controls. Another part of work is system commissioning and the qualification of facilities once design, engineering, and construction is complete.

"Based on my role for any given project, my responsibilities may change," says Yang. "For instance, I may function as a Project Engineer on one project, but as a Process Engineer for another."

"For instance, when I function as a Project Engineer, I lead a multi-disciplinary team and oversee all aspects of the project, including schedule, resources, budget for process engineering, automation and control engineering, architectural disciplines, and more," she explains. "Conversely, when I function as a Process Engineer, my work typically focuses on just the process engineering aspects of the project." She notes that for the production of biopharmaceuticals, the "products" manufactured in the facility may be either the drug substance (DS) or the finished drug product (DP) itself.

"When a biopharmaceutical product is to be produced in a liquid form (for intravenous injection), you need to design the so-called 'fill-finish process' to fill the vials or syringes or bottles," she continues. "By comparison, for a small-molecule drug the active drug ingredients (API) must ultimately be blended with other ingredients and then formulated into pill, tablet or capsule form."

"The nature of my work also varies whether I begin work on a project from the very beginning, in which case I help the engineering team to deliver the scope definition and option analysis for giving business goal, and develop schedule, cost, and scope. This is called the project-initiation phase," she says.

Yang notes that once the initial phase of a complex biopharmaceutical production facility is complete, the conceptual design begins. "During the conceptual design, we often hire an external design engineering firm to develop conceptual design for the scope defined," she says. "My role is to provide the right scope of work and direction, to line up the right resources as the owner, and to provide technical guidance."

When it comes to any pharmaceutical-related production process, there are extensive qualification requirements to be met, related to both installation and operation of the facility. "For most projects, qualification efforts are carried out by specialized third-party companies that interact with the owner company or the user of the facility that is being constructed, but that requires a lot of coordination between the engineering department (my role) and the specialized qualification team," she says. "Once the qualification is done, we hand the facility over to the operations team."

SOLID ENGINEERING EXPERTISE, FOCUSED ON BIOLOGICAL PROCESSES

During her tenure at Roche/Genentech, Yang has provided extensive engineering support and has played a supervisory/project management role in numerous projects for an array of leading pharmaceutical manufacturers, such as Bayer Healthcare, Baxter, Amgen, Novartis, Eli Lilly, Wyeth, BioMarin, and others. She has been involved in both the design and construction, and the expansion or upgrading, of complex pharmaceutical production facilities. "As a subject matter expert for downstream processes, I have also investigated disposable technologies and implemented the newly selected technologies, participating in various troubleshooting efforts," says Yang.

Earlier in her career, Yang also worked as a purification area lead engineer, where her responsibilities ranged from conceptual engineering and detail engineering through equipment procurement management and construction support for a new pharmaceutical product purification facility. She also spent nearly 15 years working for Fluor Enterprises and Jacobs Engineering—both major global engineering and construction (E&C) companies.

A FORTUITOUS FORK IN THE ROAD

"When I was in school as an undergraduate in China, my focus was really straight chemical engineering, and it wasn't until the end of my master's degree program at the University of British Columbia that I took my first real biochemical engineering," she says. "It was just one course, and very fundamental, but when I started working for Jacobs Engineering, the company was building the largest facility for Genentec in Vacaville, California. One thing led to another and I ended up working on a few large projects for Jacobs, designing and engineering facilities that would produce pharmaceutical products via biotechnology-based processes, so my expertise has grown through the years."

"I always focused my studies on conventional chemical engineering training, and I feel it was interesting and challenging for me. I realized that the ChE training provided me with good, solid understanding of engineering principles," she says. "I have been fortunate to branch off and take ownership of an important piece of the industry—bioprocessing and pharmaceuticals production—but the fundamentals of chemical engineering have remained really important in all of my work," says Yang.

"Some who focus on biochemical engineering in school may get a better grounding on the biological and bioprocessing side, but they may be forced to take a tradeoff in terms of fewer courses related to heat transfer, mass transfer, and so on. If you're going to work in an engineering role, you need those solid fundamentals," she says. "It all comes down to the same engineering basics. As long as you build a very strong background, you can explore and try new things."

Yang recalls having had an "excellent" chemistry teacher in middle school. "I knew from middle school on that I wanted to be a scientist or an engineer," she says. "Chemistry was so fascinating to me—you're very fortunate if you a can get a skilled teacher to bring you through that doorway. Nowadays, it seems like too many students just avoid the subject, because they have a preconceived notion that chemistry is hard and boring."

ON LIFE AS A FEMALE ENGINEER

Yang acknowledges that as a female engineer, she has encountered some frustrating and irritating moments in her career. "When I was in school in China, that type of sexism was less visible," she says. "During my school years, the measurement metric was simple—you were evaluated based on your grades and performance in class, so it seemed much more black-and-white, but when I first came to Canada for graduate school, I noticed very few female graduate students in the Chemical Engineering Department."

"After my first year of graduation, there was a scholarship competition between chemical engineering, metallurgical engineering, and chemistry graduate students; you had to have certain GPA and a professor had to recommend you, based on the strength of your research work," Yang recalls. "I won that scholarship, and my professor was thrilled and proud—but some of my male peers in the department said (essentially) 'wow, we didn't expect you would get it,' because I was one of the few female graduate students."

"In the workplace during the engineering and construction of complex biopharmaceutical process facilities, it can be very hard—I'm both female and considered a foreigner. Sometimes people don't take you seriously. It obviously depends on who you work for (and with), but it can be very frustrating," she says.

On recent 120-million-dollar project for Genentech, Yang says: "I had huge responsibility and was very lucky my entire team came to trust me over time based on my capabilities and performance. But there were still times I had to use my mentor to find better ways to deal with someone who was not as open-minded as others."

"As a female engineer, you must be assertive," she says. "To get where I am today, I feel like I had to work much harder than my male counterparts to prove myself, and I had to take extra effort to demonstrate my knowledge and capabilities and prove my worth."

"And, of course, those who still harbor sexist attitudes must evolve," she adds. "It's no longer 'an exception to the rule'—women in engineering and in managerial roles are here to stay."

FUTURE PROSPECTS FOR CHEMICAL ENGINEERS

"Over the past five to ten years, most chemical engineering departments have begun to add 'bio' into their name—for instance, renaming the department Biochemical Engineering, Biological Engineering, or Biotechnology—so that's definitely a trend," she says. "Clearly that expertise applies to the pharmaceutical industry, but also to the development of biofuels and chemical- and plastic-production routes that are based on biological processes or renewable feedstocks, so I feel the demand for these types of graduates will be strong in the years to come, with the use of bioengineering-based processes extending to a much wider area—whether the degree has 'bio' in the title or is just a fundamental engineering degree."

"My advice to college students is this: no matter what you do, you cannot be 100% certain, and you don't have much time to explore endlessly, because with any engineering degree you have to get into something and excel in it," says Yang. "Sometimes you need actual work experience to fine-tune the direction you want to pursue."

"And you need to go through the hard work. Kids nowadays, when it gets hard, often change focus or direction," she says. "You have to work hard; there's just no way to get around it. You have to really study your materials, spend the time, read them again and again and solve all the problems. Ask for help. Go over it again and again and again, as needed. The effort will pay off in the end." ∎

ENGINEERING & CONSTRUCTION (E&C), PROCESS TECHNOLOGY

In his own words
"Doing what you love…
Love what you do"

JAMES TURNER, PE, FLUOR CORP.

Current position:
Executive Process Director, E&C Process Technology, Fluor Corp. (Sugar Land, TX)

Education:
- BS Chemical Engineering, Texas A&M University (College Station, TX)

A few career highlights:
- Fellow of the AIChE (an honor reserved for AIChE members who have proven to be strong contributing members to the institute, society, and the chemical engineering profession)
- Received the AIChE's Fuels and Petrochemicals Service Award for distinguished service to the division and industry
- Inducted into the Texas A&M Chemical Engineering Academy of Distinguished Alumni
- Serves on the Texas A&M Chemical Engineering Deptartment Industry Advisory Council and the Texas A&M TEES Industry Advisory Council
- Leader of Fluor's Global Overpressure SME Group
- Is a licensed Professional Engineer (PE)

James Turner has more than 25 years of experience in process design for Fluor, which is a large, international engineering and construction (E&C) company. He works in the company's Energy and Chemicals Business division, which designs and constructs upstream (oil and gas exploration), downstream (petroleum refining), and chemicals-production facilities. Over the past several decades, he has worked on a wide range of projects in the petroleum-refining and gas-processing industry, in domestic and international locations.

From 2010 to 2015, Turner managed Fluor's Process Technology and Engineering group, which has roughly 150 process engineers working on hundreds of projects for various clients throughout the global chemical process industries—"ExxonMobil, Shell, Dow, Chevron, Marathon, and Ineos, to name a few," he says. Since 2015, he has worked on several projects, with particular emphasis in helping clients with the conceptual (initial planning) phase of project development.

Turner recalls that "he experienced terrible timing" when he graduated with his chemical engineering degree in 1983. "I believe it was the worst year in the last 100 years to be looking for a chemical engineering job." He recalls that fewer than 10% of his classmates had jobs upon graduation. "Many of them chose to go to graduate school, but I was tired of school and wanted to work," he says.

He took a temporary job delivering phone books to pay the bills, and by six months after graduation he was hired to join the Inside Sales Group of Teledyne Analytical Instruments, which manufactures portable and online analyzers for the process industries. The Inside Sales Group prepared proposals for prospective customers and answered technical questions that customers and sales representatives had.

"I did not consider myself particularly outgoing at the time, so working in the sales department pushed me to develop my communications and organizational skills—as opposed to my mathematical and chemical technical skills," says Turner. "I now consider it a 'blessing in disguise' that I started my career in sales—which I could not have predicted or expected a few years earlier." Teledyne "was a fast-paced, 'sink-or-swim' environment—but I was happy to find, as a young graduate, that I could succeed, and I was given a lot of responsibility and opportunity less than a year out of school," he recalls.

"After less than a year working at headquarters, I was asked to help start up a regional office in Houston—the first regional office for this company. That was in 1984—before the Internet and email were available. It was like starting a new business, and I helped develop the procedures for communicating orders and proposals back to headquarters, and shipping products and parts to customers." After a year or so, he went into outside sales. "The outside sales representatives call directly on customers and potential customers, and are typically the link between the company and the customer. I was the top salesperson for the company for a few years," he adds. He believes one of the primary reasons he was successful in this role is that his technical background, which most sales representatives did not have, gave him an advantage in understanding customers' operations.

After a few years, Turner says: "Although I considered myself successful, I decided I missed the technical challenges and decided to try to get a job in process engineering. I ended up with multiple offers and selected the offer from Fluor, where I've been for more than 25 years."

A CAREER THAT HAS EVOLVED WITH THE TIMES

Reflecting on his long tenure at Fluor, Turner notes that he's had several roles at the company, and each has tracked closely with the trends that were at work—particularly with regard to petroleum-derived fuels—to shape petroleum refineries and other operating companies throughout the chemical process industries, and the chemical engineering profession as a whole.

"Early in my career, I worked as a process engineer working on projects, where I'd basically be working on the design of the facility (mostly refinery-type projects, whatever facility we were contracted to design and build)," he recalls. "in the 1990s, there was a big push by the US Environmental Protection Agency (EPA), regulators in other countries, and many science and engineering organizations worldwide to produce so-called 'clean fuels'; these are fuels that generate a lot less sulfur and other pollutants, and thus are able to reduce pollution and emissions when they are used."

"That broad challenge required an enormous amount of technological innovation throughout the entire chemical engineering community, and demanded a lot of changes to be made to existing petroleum-refinery processes," says Turner. "I became very knowledgeable on the various processes that were required to produce the clean fuels everyone was demanding."

Consider the case for diesel fuel. The acceptable sulfur specifications for diesel were historically in the range of 5,000 parts per million (ppm). But in the early 1990s, the regulatory requirements were changed to suddenly call for maximum sulfur levels in diesel fuel to be just 500 ppm. "Even though not too many refineries had been producing diesel with sulfur levels that low, the chemical engineering community already understood that hydrotreating processes were available to reach those new, lower-sulfur objectives, so I worked on implementing some of those projects early on."

That involved Fluor engineers working with all of the major licensors of competing hydrotreating process technologies and related catalysts, including Shell, Exxon, UOP, and others. "Through these efforts, we learned more about the critical importance of maintaining good distribution inside the high-pressure catalytic reactors," says Turner. "Because diesel and hydrogen flow down through the solid-state catalyst, ensuring good distribution of the liquid and gas inside the reactor becomes extremely important to achieve the lower levels of residual sulfur in the final diesel fuel that is produced."

"About ten years later, the specs changed again, requiring refineries to produce diesel with a maximum residual sulfur level of just 15 ppm," says Turner. "With that technological challenge, the distribution inside the catalytic hydrotreater reactor becomes critical. If you have maldistribution—where parts of the petroleum feedstock are not in contact with the catalyst for the appropriate residence time—you won't reach the sulfur-removal target."

This paradigm shift in sulfur standards required the chemical engineering community to come together to better understand many critical factors. For instance, Turner says: "In most cases, we have to remove almost all the nitrogen in the diesel to allow the catalyst to effectively remove the sulfur to the new ultra-low target levels (even if nitrogen removal is not required as a product specification)." Refiners also needed to change their practices for using the same piping to move different products, since even a small contamination would cause the diesel product to go off-specification (given how tight the specifications for allowable sulfur had become).

To meet the gasoline sulfur specification of 30 ppm, petroleum refiners needed to reduce the sulfur in the gasoline material produced in the Fluid Catalytic Cracker (FCC) units. To do this, they could install "post-treater" units to remove sulfur from the gasoline components, or "pre-treater" units to reduce the sulfur in the feed to the FCC.

"New processes were developed for 'post-treaters' that could reduce the sulfur without significantly reducing the octane rating of the material and minimize the hydrogen consumption in the unit," he says. In general, "pre-treater" units are typically more expensive than "post-treaters" units, but have the benefit of improving the yields and reducing emissions from the FCC, explains Turner, adding: "Most refiners installed post-treaters units, but I was able to work on several projects that met the requirements by installing new or revamped pre-treater units."

Similarly, Turner recalls that in the mid-2000s, "we were handling a lot of capacity-expansion projects and projects to upgrade and overhaul refinery configurations so they could handle so-called 'heavy,' non-traditional crude oil streams—traditionally undesirable crude streams that were more economical to use as the starting feedstock, but harder to process into cleaner fuels," says Turner. "At one point, I was leading Fluor's process engineering team that involved 50 process engineers working on a US$10-billion project that involved thousands of people, working across several different engineering companies, to make it all come together, to advance the entire project, and ensure smooth construction, start-up and implementation."

One major highlight in Turner's successful career with Fluor came in 2014, when the company tapped him to help a client develop the conceptual plan for a major integrated petroleum refinery and petrochemical project in the Middle East. He has been working closely with Fluor's Amsterdam office on that project (in conjunction with his regular other work projects) since then, to help it move forward. "Hopefully down the road Fluor will be a major part of this project," he says.

"It was really interesting and impressive to be part of the worldwide refining industry at a time when we were all challenged to develop these solutions—and we made them happen," says Turner. "We used our collective chemical engineering ingenuity to design solutions and make the necessary changes, and today these advances are almost 'standard operating procedure' worldwide."

PROFESSIONAL DEVELOPMENT: A LIFELONG ENDEAVOR

"I think it is important to realize that the chemical engineering curriculum is not intended to give all the information and training required to properly do any job—it provides the basic framework, but people need to constantly add the knowledge that is required to do the specific job or task they are working on," says Turner. "No one knows everything and learning is a lifelong activity."

Toward that end, Turner has remained very active in the AIChE since college. Several years ago, he was elected Fellow of AIChE—an honor bestowed upon AIChE members that have proven to be strong contributing members to the institute, society, and the chemical engineering profession. He has also served as Chair Elect, Chair, and Past Chair of the South Texas Section of the AIChE, is a Trustee on the AIChE Foundation Board, and has held several positions with the AIChE, including member of the Executive Board Programming Committee, and several positions with the the Fuels and Petrochemicals Division, including Chair of the Division, Director, Newsletter Editor, and Treasurer. In 2012, Turner received the AIChE's Fuels and Petrochemicals Service Award for distinguished service to the division and industry.

RULES OF THUMB FOR PROFESSIONAL AND PERSONAL SATISFACTION

In addition to reminding others to always remain flexible and adaptable, Turner offers the following proven "rules of thumb" for students and recent graduates who are setting out to develop their careers.

- *Develop a reputation for solving the most difficult problems.* "Lots of people can solve the simple problems—only a few can solve the most difficult," he says.
- *Find something important to be the expert in, but strive to be good at all parts of your position.* Turner says: "We all have many traits that combine to be our total value to an organization; poor skills in any one area often cancel out good skills in several areas, in other people's eyes."
- *Willingly pass on your knowledge to others in your group.* "By sharing, through good team work, and by mentoring opportunities that arise, you increase your value to the group, and you will gain friends and admirers," he says.
- *Learn what others in the organization do, and what is most important to them.* "When possible and practical, take actions to make other people's job easier," says Turner. "You'll both appreciate it, and you will likely be rewarded for it down the road."
- *Continue to work on developing those skills—the ability to innovate, strong situational awareness, a healthy skepticism toward information from others, and strong judgment and decision-making skills—that will be important for the next position that you want.* "It's important to always be looking ahead," he says.
- *Position yourself so that you provide more value to your employer or client than what you are paid.* "Wise advice for students at every age," he adds. ∎

LEGAL/LAW

In her own words
"Exploring the natural synergy
between engineering and the legal profession"

ANITA E. OSBORNE-LEE, JD, ATTORNEY-AT-LAW (PRIVATE PRACTICE)

Current position:
Attorney and Counselor at Law; Registered Patent Attorney

Education:
- BS Engineering Sciences, with specialization in Computer Applications, University of Texas
- JD from Texas A&M School of Law (formerly Texas Wesleyan School of Law)

A few career highlights:
- Passing the Patent Bar Exam
- Being admitted to practice before The Supreme Court of the United States
- Persevering and thriving in a second career that is as challenging and gratifying as the first

Anita E. Osborne-Lee is a Houston-based practising attorney, and an engineer by training. She is a registered patent attorney with an interest in primarily doing intellectual property work, but, as a solo practitioner, she handles many different types of legal cases. "For the past 18 years, I have worked on whatever types of legal cases come through the door, so in addition to patent-related cases, I have handled cases involving elder-care law, wills and estates, probate, bankruptcy, guardianships, family law, and more. It's part of my psyche—I like learning new things and helping people, and I have done a lot of *pro bono* work to help others in the community."

Osborne-Lee earned her BS in Engineering Sciences from the University of Texas, Austin and, following her graduation, she spent 13 years working at the US Government's Oak Ridge National Laboratory complex system.

She and her husband Irvin Osborne-Lee—who is profiled in Chapter 7—both joined ORNL following graduation from the University of Texas (she with her BS joining the Computing and Telecommunications Division, and he with his PhD joining the Chemical Technology Division). "He always knew he wanted to ultimately teach at a university, and I always knew I ultimately wanted to be a lawyer," she says, adding: "I studied engineering as an undergraduate to help pay for law school later."

STARTING OUT

As an undergraduate, Osborne-Lee initially pursued engineering sciences with an emphasis on computing technology because, she says, "it was the closest hybrid major that really piqued my interest." She had enjoyed physics and had done very well on her SAT exam in high school.

"When I got into chemistry in college, it was a challenge—but I like a challenge," she says. "I ended up getting into chemical engineering by my sophomore year, but ultimately switched my major back to what I had originally started."

When Osborne-Lee joined ORNL after graduation, she first held a position in the Computing Technology Division, as a computer analyst of applications. "At the time I joined, ORNL was in the midst of doing a big conversion from older to newer, cutting-edge computer systems, so my combination of engineering and computer science expertise made me very useful to help that transition, and I really enjoyed the work."

After several years in that role, she transitioned to the Procurement Division that served ORNL and two other plants. There, she helped the division to modernize, automate, and upgrade the many processes and computer-related systems it used to manage data, generate reports, and manage requests-for-proposals (RFPs) for goods and services procured by the contractor(s) operating the labs on behalf of the US Government from various vendors or agencies.

PURSUING HER DREAM OF BECOMING AN ATTORNEY

After spending 13 years with the labs in several roles and jobs, Osborne-Lee decided to pursue her dream to become an attorney and to get back to Texas. She only applied to law schools in Texas, even turning down scholarships to other law schools that were offered elsewhere based on her LSAT scores. During her first year of law school, she commuted back and forth every few weeks—"with books in my book bag all the time," she says—splitting her time between Texas and Tennessee. Eventually, she and Irvin decided to concentrate their lives in Texas.

"We discussed the options and made a plan that involved a few sacrifices—because we're great planners and we had some long-term objectives in mind. During that next period in our lives, I was in Fort Worth (in law school) and he was in Houston (where he had become a chemical engineering professor at Prairie View A&M University). I was still commuting back and forth between law school and Houston, where Irvin and the children were. During Spring Break, my kids would get on a plane and come to college to stay with their Mom in the dormitory, and I'd joke with them, saying 'your Mom's your first college roommate!'"

She credits her engineering degree and her work experience at ORNL with making her time in law school a great success. "I was on Law Review during law school, and had a strong work ethic; I knew how to show up and get the work done." Once she began work as an attorney, she says "I had been a practising engineer with a strong grounding in engineering, computer science, and more, but I was not interested in being on the 'Partner track' at a law firm, so I decided to work on my own as a sole practitioner."

When she was working to build her private practice, she became affiliated with the Houston Northwest Bar Association, "and the president at the time was also a former engineer, so he understood and helped smooth my transition to law."

"I started going to those local Bar Association meetings to learn more about the law and how the court system worked, meet the local judges," says Osborne-Lee. After a few years, she became President of that local Bar Association organization. "That's pretty much how I've always been with every organization or affiliation I've had—I'll jump in for any job, any task—and such effort tends to pay off over time."

FOCUSING ON PATENT LAW AND INTELLECTUAL PROPERTY

Osborne-Lee enjoys the legal services she provides related to patent law and intellectual property (IP). "Being in the Houston area, I see many clients who work or worked in the oil industry, but now they want to do something else or retire early, and they say 'I have this great idea for an invention and I want to patent it,'" she says.

"Most people do not have a clear understanding of what a patent or any IP is, and there's a big difference between a natural progression of improvement on an existing invention versus an entirely new invention," she says. "I have had so many conversations to help people understand which ideas or inventions are patentable, whether their idea can or cannot be patented based on their employer, and more. And, patent searches often reveal the idea has already been patented but is still not making any money for the inventor."

BLENDING ENGINEERING WITH LAW

"The law is black-and-white-and-grey and, being a trained engineer, I tend to think in terms of black-and-white," says Osborne-Lee. "Sometimes (in legal proceedings), there's not always 'one answer'; rather, the best answer is the one that has the best argument, based on the assumptions used to build the argument, the facts, history, minutiae, etc."

However, in trial court with an appellate system, you typically have to challenge, change, or expand an earlier ruling by using a stronger argument or better facts," she says. "That approach draws great similarities to the realm of engineering, where you discover and improve things by building on the rules of science and engineering."

"As a trained engineer, I have a whole other lens through which to see the world, and I have garnered respect from my peers, professors, and colleagues from the time I enrolled in law school with an engineering undergraduate degree to the many cases I work on today," she continues.

During law school, Osborne-Lee recalls helping and tutoring her younger peers, who were struggling with the workload or complexity of the courses. She used to joke with them, saying: "I gave up my engineering career and left my family to be here in law school, so you need to stop complaining and just get busy."

ON MENTORING AND BEING A ROLE MODEL

During the years that she and Irvin were both working for ORNL, they both got involved with a program called Science in Action, which was carried out in several local schools as a way to inform and inspire young students to become more familiar with the sciences and engineering.

She has also been active in the National Organization for the Professional Advancement of Black Chemists and Black Chemical Engineers (NOBCCHE), serving as National Secretary of the organization. "I've been particularly interested in encouraging and mentoring black females in engineering," says Osborne-Lee.

Meanwhile, Osborne-Lee says that she has both experienced and fought against racism and sexism at times in her career. "I was born black, female, and intelligent, with educated parents, and I've always been outspoken. I always ask questions and challenge what I see that's not fair," she says.

"You'll always encounter certain people who will never accept a female or a black boss, based on their own ignorance or prejudices," says Osborne-Lee. "You have to be resilient. I find great joy whenever I have proven I could succeed with grace and garner respect from my peers in the scientific or legal community; I knew the next generation would have it easier than I did."

"I've made progress in places where people never thought I could," says Osborne-Lee. "It's nice to have money, professional stature, and recognition—but to have someone dislike you and dislike everything about you based on their own ignorance but still respect you is worth more than a million dollars." ∎

PROCESS SAFETY CONSULTING

In his own words
"Using a lifetime of metallurgical and chemical engineering
expertise to prevent catastrophic events"

STEVEN D. EMERSON, PhD, PE, Emerson Technical Analysis, LLC

Current position:
Principal, Emerson Technical Analysis, LLC (Corpus Christi, TX)

Education:
- BSc Metallurgical Engineering, Colorado School of Mines (Golden, CO)
- Master of Engineering Administration, Bradley University (Peoria, IL)
- PhD in Chemical Metallurgy, University of Arizona (Tucson, AZ)

A few career highlights:
- Fellow of the American Institute of Chemical Engineers (AIChE)
- Technical Advisory Committee, Mary Kay O'Connor Process Safety Center, Chemical Engineering Department, Texas A&M University (College Station, TX)
- Is a licensed Professional Engineer (PE)

Through his consultancy, Emerson Technical Analysis, LLC, Steve Emerson specializes in process safety, with a particular focus on preventing catastrophic events, "and it's been exceptionally fun," he says, enthusiastically. "Working as an independent consultant allows me to use the wisdom, creativity, and expertise I have amassed through my prior experience in the chemical process industries and apply my efforts to a very worthwhile goal—preventing catastrophic process safety incidents."

Emerson notes that his challenging and interesting career has not unfolded in a straight line. "It's had many 'zigs and zags' along the way," he says. "Who could ever have imagined a young farm kid who was raising pigs in Colorado going on to have a fascinating, enriching engineering career and advising clients all over the world?"

Emerson studied metallurgical engineering in college, and notes that there is plenty of cross-over between this discipline and chemical engineering.

In his first job following graduation, Emerson worked in a steel mill, and that helped him to both hone and apply the metallurgical and chemical engineering expertise he learned in school—including fundamental concepts of transport phenomenon, mass transfer, heat transfer, and more—in unique and unusual ways. "I worked all shifts around the clock, which gave me a chance to see what really happens—not just what the boss thinks takes place—and it was an exceptionally eye-opening experience for a young engineer," he recalls fondly.

Recalling his PhD dissertation, which was focused on optimization of electric-furnace steelmaking, Emerson says: "I got to see the beauty of the technology, and apply concepts beyond where they had already been shown to apply." He continues: "Engineers are capable of doing that. There's a creative streak in doing this and, for a farm kid like me, that's been immensely satisfying."

For several years earlier in his career, Emerson worked for Kerr-McGee Corp., the worldwide petroleum exploration, refining, and chemicals company, where he was responsible for helping the chemical giant to shape its worldwide objectives and policies related to safety, environmental protection, and loss prevention. These efforts included developing environmental-compliance programs, directing regulatory-compliance auditing and emergency-preparedness, and ensuring proper engineering design related to process safety. Emerson says: "During my time at Kerr-McGee, we cut the worldwide lost-time accident rate by 50%, and reduced the all-injury rate well beyond 50%," he says, proudly.

ENGINEERING CONSULTING, IN TWO DISTINCT FORMS

For the most recent 20-or-so years, through his consultancy, Emerson's work has been divided into two buckets: he provides engineering consulting and advisory services to operating companies throughout the chemical process industries, and he provides engineering analysis needed to support litigation (for clients on both plaintiff and defense sides). Through this type of legal work, Emerson explains that he "carries out investigations of tragic events that have occurred and digs more deeply into the root causes that allowed such catastrophic events to take place."

"Such an analysis is always a very, very interesting and rewarding exercise," he notes. "There are often common themes running through these types of safety incidents—such as shortcuts that were taken, or a lack of sufficient effort or investments made in safety-related systems. Of course, those insights prove very helpful in supporting the prevention work with operating companies."

AN ENTREPRENEURIAL OPPORTUNITY ARISES

Twenty years after his PhD program, he was one of two former students approached by his graduate advisor to help found a start-up company, Qualitech Steel Corp. Emerson was to be responsible for an iron carbide facility demonstrating exciting new process technology. "Our goal was to perfect the technology, and we secured US$500 million of capital funding for both a steel mill (outside Indianapolis, Indiana) and the iron carbide facility (in Corpus Christi, TX), and fast-tracked the development of the iron carbide complex, going from financing to start-up in just 24 months," he recalls.

"It was a thrilling time to see a US$125-million greenfield chemical plant come online to convert iron oxide (Fe_2O_3) to iron carbide (FeC_3) using a revolutionary process," says Emerson.

Unfortunately, the plants came online in late 1997—"exactly the wrong time in the steel industry," says Emerson. "We came up against a huge influx of Eastern European steel, and it flooded the US market and drove prices far below the break-even prices for US production to be viable, and Brazil flooded the market with briquettes that drove the iron carbide business out of business."

Eventually, the company was forced into Chapter 11 bankruptcy. "I lost a lot of money on that deal but it was an extraordinary experience I would not trade for anything," he says.

LOOKING BACK TO HIGH SCHOOL

Emerson went to a small high school in northern Colorado, and was the third generation to be raised on his family's farm. "Because we were growing up on a farm, all of my siblings and I learned to raise livestock. I raised pigs, and I sold them to help pay for my tuition at the Colorado School of Mines" he says.

He said he always knew engineering would be a "challenging but difficult" field, but he had always been good at math and science and technical drawing. "I'm proud to say, I had a bit of a rocky start but ended up a Dean's List student," he adds.

"For me, my engineering degrees were both the ticket and the prerequisite to accomplish what I wanted to do and, from the very start, I was entrusted with supervisory roles," says Emerson. "My degrees gave me the credibility to accomplish and pursue my ambitions, to work with teams, to launch businesses, and eventually to be an expert witness in a variety of very challenging technical fields. I couldn't have dreamed for more."

"I enjoyed understanding statics and dynamics, thermodynamics and kinetics," recalls Emerson. "I've found that—with the fundamental understanding of those concepts—you can apply the intrinsic logic to so many things you will encounter in your life, and you can understand how things operate and then take appropriate action."

ADVICE FOR TODAY'S STUDENTS

Thinking of today's engineering students, Emerson offers this advice: "There are going to be times when enthusiasm starts to wane and pretty significant difficulties arise. It will be challenging, but remember that when you're in an engineering discipline, everyone is pretty much in the same grind."

He admits that he "really had to work at it, and was reluctant to admit his failings or go engage with the professors." However, he recalls: "When I did, I found that my professors sincerely wanted to help elucidate the concepts, and did want to see success in their students. That was really a breakthrough for me, because I think my natural tendency is to not admit failure or ask for help." To all students, he says emphatically: "Take advantage of all the support that is being offered." ■

REFERENCES

Center for Chemical Process Safety. 2007. Guidelines for Risk Based Process Safety. Hoboken, NJ: American Institute of Chemical Engineers and John Wiley & Sons, Inc.

Chemical Engineering Progress. 2017. Profile, Innovating Plant-Based Vaccines and Therapeutics. April, 29–34.

Levesque-Tremblay, Gabriel and Darlene Schuster. April 2017. SBE Update-The Blossoming Field of Plant Synthetic Biology. *Chemical Engineering Progress,* 28 (4), 3–20.

Linsenmeier, Robert A. and David W. Gatchell. 2008. What Makes a Bioengineer and a Biotechnologist? In Madhavan et al.

Madhavan, Guruprawsad, Barbara Oakley, and Luis Kun (Eds). 2008.*Career Development in Bioengineering and Biotechnology,* New York: Springer Science+Business Media, LLC.

Rubin, Edward S. and Cliff I. Davidson. 2001. Introduction to Engineering and the Environment. New York: McGraw-Hill, Inc.

Shelley, Suzanne A. 2005. Nanotechnology: Turning Basic Science into Reality, in Louis Theodore and Robert Kunz. *Nanotechnology: Environmental Implications and Solutions.* Hoboken, NJ: Wiley-Interscience

Shih, Patrick M. 2017. A Brief History and Outlook on Plant Engineering. *Chemical Engineering Progress,* 28 (4), 24–25.

Website of the Project Management Institute, www.pmi.org/.

Future Directions in Chemical and Biomolecular Engineering: A Few Examples

Imagining in excited reverie that the future years had come.

William Butler Yeats (1865–1939)

CONCEPTION, DESIGN, CONSTRUCTION, OPERATION, AND MAINTENANCE OF SUSTAINABLE LIFE-SUPPORT SYSTEMS FOR BASES ON THE MOON AND ON MARS

A *life-support system* for humans provides for supplies of oxygen, water, food, and a temperate environment, as well as management of wastes and carbon dioxide. It also provides for other essential amenities.

A sustainable system must operate indefinitely without failure or compromise.

A sustainable life-support system must be able to provide life-sustaining capabilities for the user, indefinitely, without failure or compromise. For instance, to sustain human life in another atmosphere, the ability to recycle carbon dioxide that is produced during breathing and use it to produce oxygen at the appropriate concentrations of both gases in an enclosed system is fundamental. The proper disposal of wastes, preferably using purification and recycling techniques, and providing for maintenance of a water balance in the enclosed space are also essential. The buildup of odors and other undesirable compounds would need to be prevented. These are just a few examples of the fundamental applications of chemical and biomolecular engineering in the creation of a sustainable life-support system to advance space travel and to enable humans to work and live in confined spaces or oxygen-poor environments (see Figure 10.1).

Within the realm of space travel and efforts to maintain life on other planets that are inhospitable to human inhabitation, it is also necessary that the enclosure remain intact and impermeable even though there is no atmosphere on the moon to shield it from numerous small meteors. It is not clear that the thin atmosphere (which is rich in carbon dioxide but devoid of oxygen) on Mars would be nearly as effective in burning up incoming meteors compared to the earth's denser and highly oxidizing atmosphere.

Ease of repair of the system would also be essential, and the ability to create and maintain the system using components that would be small enough to allow them to be launched into space from the earth would be needed. Note that delivery of these spare parts from the earth would not occur in a short period of time. To be sustainable over the long run, the ability to make onsite repairs could be the preferred option for a limited number of parts of the system.

NASA (National Aeronautics and Space Administration), ESA (European Space Agency), RFSA (Russian Federal Space Agency), and CNSA (Chinese National Space Administration) have been actively researching sustainable life-support systems, starting even well before the International Space Station, with systems like Skylab. But much more needs to be done to achieve

FIGURE 10.1　Chemical and biomolecular engineers have countless roles to play in developing workable solutions to support modern life. And many of them also dream big, putting their engineering ingenuity to work creating solutions to address futuristic challenges as well. Shown here is a drawing of an imagined base to support life on Mars. (Source: Shutterstock ID 416044723)

true sustainability, and chemical and biomolecular engineers will certainly have many roles to play to address and solve these future-looking challenges.

DEVELOPMENT OF NEW PROCESSES FOR THE ECONOMIC RECOVERY AND RECYCLING OF ELEMENTS IN ELECTRONIC WASTE

Electronic waste is a serious problem. In 2014, approximately 41.8 million tons of electronic waste was generated worldwide (see Figure 10.2).

Electronic waste (e-waste) includes discarded television sets, monitors, computers, printers, modems, cell phones, microwaves, radios, batteries, and lamps.

Although nearly all e-waste is recyclable, only about 6.5 million tons of total global e-waste generation in 2014 was treated by national "electronic take-back" systems; the rest is typically discarded and ends up in a landfill. Recycling efforts can recover valuable constituents such as plastics, glass, and metals. Reuse of certain engineered components is also feasible. The rate of recycling of e-waste in the United States was 29 percent in 2012, 24.9 percent in 2011, and 19.6 percent in 2010 (Levin 2017).

Citizens of the United States discard roughly 100 million cell phones every year and 112,000 computers every day either by recycling or disposing of them in landfills and incinerators. Improperly constructed landfills can lead to groundwater contamination, potentially affecting drinking water supplies.

With proper recycling efforts, one million cell phones could lead to the recovery of 20,000 pounds of copper, 20 pounds of palladium, 550 pounds of silver, and 50 pounds of gold. Cell phones contain very high amounts of precious metals such as silver and gold, and with the current sub-optimal recycling efforts, Americans throw away approximately US\$60 million worth of silver and gold per year.

The US EPA reports that e-waste currently accounts for about 2 percent of the solid waste stream, but that amount accounts for 70 percent of hazardous waste that is deposited in landfills. Importantly, electronic devices are a complex mixture of several hundred materials, including the rare earth elements. Thus, e-waste is rich in many toxic heavy metals, such as lead, mercury,

FIGURE 10.2   Our modern life creates countless waste streams, which, if not properly managed, can create lasting environmental damage. Chemical engineering know how is routinely put to use in the quest to create and commercialize processes that can help break down complex waste streams, such as the electronic waste streams that are created by society's heavy dependence on digital devices, computers, and gadgets. (Source: Shutterstock ID 606226475)

cadmium, chromium, arsenic, and beryllium, which should not be released into the environment when the device is no longer needed.

Methods for the recovery and recycling of the various metals in e-waste have been developed, but there is much work still to be done to increase the effectiveness, infrastructure, and overall economics of these environmentally beneficial practices. Chemical engineering and chemistry are the two best disciplines to foster ongoing breakthroughs in new and better ways to recycle the components of e-waste.

INVENTION AND MASS PRODUCTION OF LOW-COST AND DURABLE SOLAR CELLS

Photovoltaic or solar cells are devices that convert light into electricity via the photovoltaic effect.

Sunlight is free and available almost everywhere, so solar energy provides a promising, desirable renewable energy source, and an important alternative to fossil energy sources such as coal and petroleum. The potential advantages of solar energy have led to extensive research and development, and to date there are more than 20 types of photovoltaic or solar cells currently available (see Figure 10.3).

Chemical engineers have played a vital role in this research-and-development area, yet there is much more work to be done. In addition to research on new types of photovoltaic cells, there are opportunities to improve the existing processes that are used to manufacture solar cells. Innovation in the design and operation of existing and new manufacturing processes provides another great chemical engineering opportunity.

DEMONSTRATION AND COMMERCIALIZATION OF PRACTICAL METHODS TO CAPTURE AND SEQUESTER GREENHOUSE GASES

Greenhouse gases are gases that promote global warming by absorbing electromagnetic radiation (such as sunlight and infrared heat rays) in the infrared portion of the electromagnetic spectrum. Greenhouse gases include carbon dioxide, methane, sulfur hexafluoride, and many fluorinated and chlorinated hydrocarbons that have been used as refrigerants.

INDUSTRY VOICES

"I hope you can find an experienced mentor who can encourage you to set your sights higher and higher—to exceed your best expectations in your career and life. Only you can set your own limits, so think big."

—Roy E. Sanders,
PPG Industries (Retired)
and Process Safety
Consultant (Current),
page 178

FIGURE 10.3   Chemical engineers bring their expertise to bear during the ongoing quest to create improved designs to harness the sun's energy and convert it into electricity via photovoltaic cells and solar panels. This photo shows an aerial view of solar energy farm in Qinghai province, China. (Source: Shutterstock ID 63565958)

Have you ever noticed that the air inside a greenhouse is warmer than the air outside a greenhouse? This difference is greatest on sunny days. This is because the glass windows of the greenhouse allow visible light to easily pass through the windows. That solar radiation is converted to heat when the contents of the greenhouse absorb it. The heat is converted to infrared radiation, which is blocked by the glass windows, keeping the heat inside. Gases in the atmosphere that absorb infrared radiation are called greenhouse gases because they increase the temperature of the earth in a manner like the glass windows of a greenhouse (see Figure 10.4).

The absorption by the earth of solar energy is converted to heat. This heat, which is radiated from the warm earth, is radiated as infrared radiation. This infrared radiation is absorbed by greenhouse gases in the atmosphere, thus increasing the atmospheric temperature of the earth. If our world did not have an atmosphere to capture solar energy, the temperature at the surface of the earth would be $-18°C$.

In the last several decades, the atmospheric concentration of carbon dioxide and other greenhouse gases in the atmosphere above the earth has risen continuously, contributing to global warming. Consequently, chemical engineers have been developing methods to address this issue by capturing carbon dioxide in stack gases from electric power plants (to prevent it from entering the atmosphere). Chemical engineers and chemists have also been reducing atmospheric emissions of other greenhouse gases.

Furthermore, because natural gas produces less carbon dioxide per unit of energy produced than coal or oil, many power plants are converting their power-generation systems to use natural gas as the preferred fuel source instead of traditional coal. Meanwhile, newly recoverable and relatively inexpensive shale gas (natural gas that has been recovered from underground shale deposits by using new types of hydraulic fracturing of these underground formations) has also been a major factor in in bringing natural gas, an "environmentally better" fossil fuel, into greater use.

Methods under development for the removal of carbon dioxide from stack gases include absorption with amines or other absorbents. In 2017, the US Department of Energy (DOE) partnered with a commercial utility to start up the first commercial-scale carbon dioxide absorber on

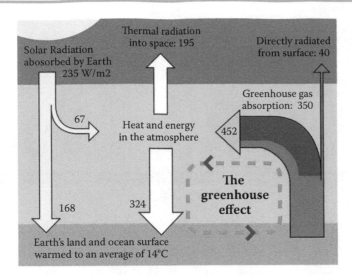

FIGURE 10.4   Greenhouse gases include carbon dioxide, methane, and sulfur hexafluoride, as well as many fluorinated and chlorinated hydrocarbons that have been used as refrigerants. A buildup of these gases impacts how energy flows in the earth's atmosphere (via a process called the greenhouse effect). Because chemical and biomolecular engineers work throughout the many industries whose activities contribute greenhouse gases to the atmosphere, their expertise is in strong demand to continue to imagine, devise, and implement viable, cost-effective strategies to minimize the amount of greenhouse gases that can reach the atmosphere, and devise strategies to counter the effects of such buildup in the atmosphere. (Source: Greenhouse Effect, Wikipedia, contributed by user Rugby471; used under GNU 1.2)

an existing US electric power plant. The captured carbon dioxide is being sent by pipeline to an existing oil field. There, the carbon dioxide is being injected deep down into the underground oil formation to recover more of the oil left behind in the reservoir after conventional oil recovery.

Ultimately, the best long-term solution for reducing emissions of greenhouse gases is to replace the processes that generate them in the first place with alternative processes that do not produce greenhouse gases at all (or at least produce them in greatly reduced amounts)—for example, by replacing fossil-fuel-fired power plants with engineered facilities that can produce electricity from wind energy and solar energy.

Chemical engineers have made major contributions to decreases in the production of greenhouse gases, and clearly there is more work to be done, so this area will continue to provide a range of exciting career pathways for chemical engineering graduates.

If society decides to invest in carbon dioxide capture and sequestration, then carbonation is one potential means of storage.

Carbonation is when pressurized carbon dioxide gas is dissolved in a solution.

A recent engineering conference held in March 2018 (CEI 2018) was focused on promoting global research and development on accelerated carbonation.
The sixth edition of the Accelerated Carbonation for Environmental and Material Engineer (ACEME) Conference aimed to promote research and development activities on accelerated carbonation and mineral carbonation in the context of carbon capture, utilization, and storage (CCUS).

Relevant topics include:

- Principles and kinetics of accelerated carbonation
- CO_2 capture and storage by mineral carbonation
- Accelerated carbonation of alkaline materials including industrial wastes, lime, cement, and concrete
- Utilization of the carbonated materials
- Pilot- and full-scale applications

FIGURE 10.5   Shown here is a technician sampling *Alga spirulina* from a captive algae farm. Algae farms are being used to grow and harvest various forms of algae to capture carbon dioxide, produce oxygen, and yield a biomass product that can then be used for other purposes. (Source: Shutterstock ID 426386110)

DEVELOPMENT OF NEW METHODS FOR THE EXTRACTION AND PURIFICATION OF WATER FROM SALINE SOURCES AND WASTEWATER STREAMS

The effective and inexpensive purification of water remains a major objective for agricultural, domestic, and industrial uses. As discussed earlier, the removal of salt and other materials from saline water resources and wastewaters requires energy. Different processes require different amounts of energy. For example, distillation of water requires more energy than reverse osmosis, especially for relatively dilute saltwater solutions. But distillation may ultimately remove more contaminants than reverse osmosis.

The degree of purification of water that is ultimately required depends strongly on the intended use of the water. For example, water that is produced for irrigation requires much less purification than potable water; water that is intended for use in microelectronics requires much greater purification than potable water.

One potential new water-purification technique would be to use greenhouses containing algae farms (see Figure 10.5). Designed with inclined planar windows, such greenhouse structures would allow condensed moisture that had been collected on the windows to flow downward into troughs around the perimeter. Carbon dioxide from cooled power plant fluegases could be injected into the greenhouses to provide photosynthesis by the algae. The captive algae farms could be used to produce oxygen and plant biomass for unconventional energy production.

As discussed in Chapter 2, new and better separation technology options for the purification of water have been developed with important contributions by chemical engineers. However, further research, development, and design innovation by chemical engineers is needed to develop improved designs that can produce desired separations even more efficiently and cost-effectively.

DEVELOPMENT OF IMPROVED HEMODIALYSIS MACHINES

Kidney failure is a major medical condition worldwide that requires substantial efforts to treat or cure. Because healthy people have two healthy kidneys, it is sometimes possible for a kidney donor to give one healthy kidney to a patient whose two kidneys have failed. However, transplant surgery is costly and brings potential risks, and suitable donor matches are not always available.

INDUSTRY VOICES

"Throughout my career, I have focused on process development and that has allowed me to work on very different and interesting things. It's gratifying to feel you're doing something that is making a real difference on issues in energy and chemistry that are impacting everyone—issues that are making front-page news."

—Joseph B. Powell, Shell, page 181

Therefore, kidney dialysis is widely used to help remove unwanted compounds from the blood of patients whose kidneys are no longer able to serve their natural function of removing those compounds from their blood stream. Patients with kidney failure require dialysis for several hours, several times a week.

Dialysis is a separation technique or unit operation that employs a semipermeable membrane between two solutions to allow for the selective transfer of specific compounds across the membrane by molecular diffusion, while retaining other components. Kidney dialysis or hemodialysis refers to the use of specialized dialysis membranes to purify blood.

In hemodialysis, salts, urea, and other waste compounds in blood transfer across a semipermeable membrane from the patient's bloodstream, flowing through an extracorporeal shunt on one side of the membrane into a receiving stream of dialysis water on the other. At the same time, the specially designed membrane prevents the loss of essential components in the bloodstream, such as immunoglobulins and red blood cells (see Figure 10.6).

Chemical and biomolecular engineers have been involved in the design, deployment, and optimization of hemodialysis systems since their inception, making many important contributions. However, the medical community is always seeking improvements in the selectivity and permeability of the engineered membranes, and in the configuration and manufacture of hemodialysis membrane systems. There are opportunities in hemodialysis for chemical and biomolecular engineers, as well as biomedical engineers, to be at the forefront of ongoing lifesaving advances in this important medical arena.

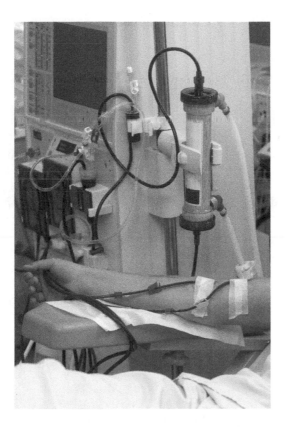

FIGURE 10.6   Kidney dialysis is used worldwide to remove unwanted salts, urea, and other compounds from the blood of patients who have severely impaired kidney function. Shown here is a typical kidney dialysis machine with electronic monitoring. The process uses specialized semipermeable membranes and is a classic example of an area in which a variety of chemical engineering inventions initially developed and still deployed in industry have also been put to use in the medical/biomedical arena. (Source: Shutterstock ID 50011900)

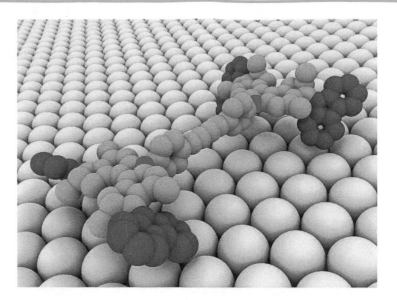

FIGURE 10.7   Increasingly, chemical engineering knowhow is being used to explore and exploit systems that have nanoscale proportions, with efforts aimed at discovering and manipulating materials at where new, advantageous characteristics and properties begin to emerge. Shown here is a model of a "single-molecule" car that can advance across a copper surface when electronically excited by an STM tip. (Source: C&EN, courtesy of Dr. Ben Feringa, https://cen.acs.org/articles/94/web/2016/10 /Molecular-machines-garner-2016-Nobel-Prize-in-Chemistry.html%7C2016)

DEMONSTRATION AND COMMERCIALIZATION OF BETTER MOLECULAR MACHINES FOR BIOMEDICAL APPLICATIONS

Molecular machines are nanoscale machines that are created by special arrangements of a molecular system to produce motion or other desired activities.

Molecular machines are single molecules that behave much like the machines people encounter every day: They have controllable movements and can perform a task with the input of energy. Examples include a tiny elevator that goes up and down with changes in pH and a super-small motor that spins in one direction when exposed to light and heat. Many in the field speculate that molecular machines could find use in computing, novel materials, and energy storage (Chemical and Engineering News 2016).

Today, scientists and engineers are creating primitive but innovative machines at the molecular level (see Figure 10.7). This is a new technology that is in its infancy, enabled by scientific advances such as *scanning tunneling microscopy* (STM) and *atomic force microscopy* (AFM). These advanced microscopic methods allow for the observation and arrangement of individual atoms and molecules. Some of these tiny machines will eventually find practical applications in medicine and human health.

Chemical engineers, bioscientists, biomolecular engineers, biomedical engineers, chemists, and physicists will continue to make important new discoveries with molecular machines.

ADVANCEMENT OF BIOMOLECULAR ENGINEERING OF MICROORGANISMS THAT CAN EFFICIENTLY SYNTHESIZE THE PROTEINS IN HUMAN BLOOD SERUM

In the past, insulin—a mid-sized polypeptide or protein composed of 51 amino acids that is used to help regulate blood-glucose levels in patients with diabetes—was extracted from the pancreas of animals in commercial slaughterhouses. Biomolecular engineers have already played an important role in creating alternative, commercial-scale processes that instead use either bacteria or yeast to synthesize insulin for use in treating patients with diabetes (see Figure 10.8).

Human insulin

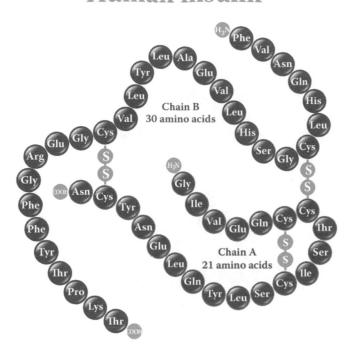

FIGURE 10.8 Human insulin, a mid-sized polypeptide or protein composed of 51 amino acids, provides an essential treatment option for patients with diabetes. Historically, it was extracted from the pancreas of animals in commercial slaughterhouses. However, biomolecular engineers have developed alternative processes for producing insulin on a commercial scale using both genetically engineered bacteria and yeast to synthesize the lifesaving product. (Source: Shutterstock ID 99098642)

Many of the proteins in human serum are larger and more complex than insulin. Would it be possible to synthesize these molecules using bacteria or yeast? Would that offer any advantage over the current source: human blood donations? The pursuit of this, along with other important biomolecules, offers potential opportunities for ongoing contributions by biomolecular engineers working in close collaboration with bioscientists.

CONCLUDING REMARKS FROM THE AUTHORS

A wise old sage once said: "Science is what is…. Engineering is what you want it to be…" The power of engineering—and chemical and biomolecular engineering in particular—is legendary in terms of its direct impact on the health, safety, and well-being of humanity and the societies in which we live and thrive.

We wrote this book to tell you about the many diverse career options and opportunities—both existing and futuristic—that are available to people who choose to major in chemical and/or biomolecular engineering, and earn a bachelor's degree, master's degree, or PhD in either discipline. Chemical and biomolecular engineers have already made many important contributions to society and will continue to make many more. As this book aims to demonstrate, such highly trained technical professionals also have many exciting industry sectors and career pathways to consider, as these two engineering disciplines are inherently useful, and highly trained graduates in both fields are inherently flexible and adaptable, growing and contributing as society's needs and discoveries continue to evolve.

If you decide to enter this field, we welcome you to the profession. If not, we hope that you will have gained an appreciation of how the practice of this profession is enabling so many aspects of the development of the technological foundation of a sustainable society.

SEMICONDUCTOR INDUSTRY

In his own words
"Combining chemical and electrical engineering expertise
to optimize semiconductor fabrication"

CHRISTOPHER BOWLES, TEXAS INSTRUMENTS

Current position:
Worldwide Reliability Engineering Manager, Semiconductor Quality Organization, Texas Instruments Inc. (Dallas, TX)

Education:
- BS in Chemical Engineering, Texas A&M University
- BS and MS in Electrical Engineering, University of Texas (Dallas)

Some career highlights:
- Elected by his engineering peers to be a Distinguished Member of the Technical Staff at Texas Instruments
- Key member of the successful start-up teams for both of TI's 300-mm wafer-fabrication facilities
- Holds several patents

Christopher Bowles is one of those rare chemical engineers who found his calling early and has remained with the same company since graduating with his BS "Throughout my professional career at Texas Instruments (TI), I have had the opportunity to take on many different and interesting roles, mostly associated with semiconductor technology development and wafer-fabrication process development, and it's been consistently challenging and rewarding."

The Reliability Engineering group that Bowles manages at TI is responsible for the qualification of products across the company's business portfolio, including both analog and embedded processing businesses serving a wide variety of customer applications. "Our group also provides consulting services related to quality and reliability, and we work with TI's customers to establish a reliability profile for TI products that are being used in specific customer applications."

Bowles has held several engineering positions that allowed him to interface directly with TI's customers. Several years ago, he was the Wafer Fab Quality Manager for TI's Richardson Fab facility—the semiconductor industry's only 300-mm fabrication facility dedicated to producing analog semiconductors. He views his work as a key member of the start-up team for this facility—which came online on schedule—as a major career achievement.

THE ROLE OF CHEMICAL ENGINEERS IN SEMICONDUCTORS

Semiconductor fabrication relies on many chemical processes, including chemical vapor deposition, plasma etching, ion implantation, chemical cleaning, furnace oxidation and annealing, and more. "Having a great fundamental grasp of chemical engineering principles—such as chemical processes and reactions, fluid transport, heat transfer, mass transfer, reactor design, thermodynamics, and more—allows for detailed understanding of how these processes work," says Bowles. "Chemical engineers," he continues, "are able to devise ways to achieve desired outcomes in a stable manner, to recognize what potential problems exist, and to understand how certain issues might contribute to poor quality or reliability."

A few years into his tenure at Texas Instruments, Bowles began to ask himself "what else do I need to do to maximize my opportunities at this company and in the semiconductor-fabrication space?" To expand his expertise and broaden his career opportunities within the company and the industry, he began taking classes in electrical engineering at the University of Texas (Dallas).

He eventually earned two more degrees—a BS and an MS in Electrical Engineering—while still employed full-time by TI. This effort has paid off and has proven to be a powerful combination, helping him to "effectively bridge the gap between manufacturing processes and the electrical performance of our products," he says.

"The classes dovetailed so nicely with the work I was doing at TI that I ended up earning both a bachelor's degree and master's degree in Electrical Engineering, and this combination has been crucial to my success here," he adds.

REFLECTING ON HOW HE GOT HERE

"Given my strengths and what I enjoy doing, I know that this was the right choice for me. I don't just think of engineering as *my profession*—I think of it as *who I am*," he muses. "I *am* an engineer. It's a major aspect of my personality, and a major aspect of my orientation toward life and the inherent problems of life."

"There was never much question in my mind that I would pursue some form of engineering," he recalls. "The way my mind works naturally put me on the path toward engineering—I was always tinkering with things when I was growing up, but the fact that my Dad was engineer made it seem not at all scary," he says.

Bowles' father was an electrical engineer at Texas Instruments. "He'd invite me to come to work with him a few times each semester, so I got to see what he was working on, and to see that he was highly regarded in his workplace," says Bowles. "He was practical, inventive, innovative, and a great problem-solver. I admired his abilities and wanted to be just like him."

"In high school, I enjoyed chemistry and did well in both regular chemistry and AP chemistry, so this was a big factor in pushing me toward chemical engineering," he recalls. Compounding his interest in this course of study, Bowles recalls: "When I was a senior in high school, I received in the mail a magazine that profiled various college degrees, including average starting salaries. Chemical engineering had the top average starting salary among all of the degrees' profiles, so that helped to clinch my decision!"

Bowles graduated from Texas A&M in 1984—at a time when, he says, the chemical engineering curriculum was almost exclusively dedicated to the oil-and-gas and chemical industries, and both of these industries were in a downturn. "There was little exposure to other application areas, such as microelectronics fabrication. Nonetheless, my fundamental chemical engineering education was sound and with that flexible degree I have been able to create a very successful career in the semiconductor industry."

Starting out in the semiconductor space, Bowles says, with a laugh: "I fully expected that I would spend a few years working in semiconductors, and would then get a 'real' chemical engineering job in the process chemicals industry." Instead, he says, "I happened into a fascinating, challenging, fulfilling career at TI, and I've always appreciated the people I've worked with here and the integrity of the company, which aligns well with my values."

My career is nothing like what I imagined it would be—but I have never been disappointed in my reality versus my imagination."

THE POWER OF TEAMWORK

Reflecting on his own career experience, Bowles says that chemical engineers should not underestimate the importance of collaboration. "Nobody ever has an engineering career in a vacuum. Your ability to collaborate with a broad cross-section of people to execute as a team will differentiate you."

And when it comes to engineering teamwork, "diversity is a strength," says Bowles. "People who come from different backgrounds will bring different perspectives and approaches to any endeavor. You do best when you leverage all of these differences for team success."

While waxing poetic, Bowles adds: "You've got to lead courageously and be willing to take risks and push for what's right, based on the best available data." However, he also notes that there's a fine line between courageously taking a position and being stubborn. "While you should courageously take a position, don't get hung up on always being right or getting your way. When your position turns out not to be the best one, you've got to admit it and move on—for the sake of the team and your own integrity and reputation."

"Being stubborn used to be my Achilles' heel, and for a long time it was very difficult for others to pull me off of my ideas," says Bowles. "But within a trusted, collaborative work space, and with lots of learning and hard-earned wisdom, I'm able to listen to, and evaluate, others' ideas much better." In the past 15 years, "I've never told anybody 'you're wrong and I'm right!'" he jokes. "Sometimes you pay for being stubborn—that was a hard-won lesson for me."

Meanwhile, Bowles also underscores the importance of mentors—if you can find the right ones, who are willing to provide constructive feedback (even proverbial "tough love"). "You won't get much growth out of the relationship if the mentor just tells you how great you are doing," says Bowles. "If your mentor isn't giving you constructive help, or if you aren't willing to work with it, you're both missing the point."

CHANGE IS THE ONLY CONSTANT

"I have been surprised by turns of events many times during my career, and I could never have forecast 32 years ago what the world would be like now and where my career would go as a result," says Bowles.

While he is grateful to have worked for Texas Instruments in many different areas of the semiconductor business, Bowles notes that most professionals will lose their job at some point in their careers, due to layoffs, corporate restructuring, and so on. "I have seen such upheaval kill the careers of some people, but I have also seen others benefit from ending up in better positions, and in more interesting or rewarding career tracks, thanks to the change that was forced upon them." The key difference, he notes, "seems to be the attitude and pro-activity of the person losing their job. Those who cannot adapt will be left behind."

Toward that end, he also reflects that successful engineers never pass up the opportunity for rigorous self-examination—"so they are very aware of their own strengths, weakness and interests," he says. And they never pass up ongoing opportunities to broaden their expertise and experience—"so they remain valuable in the changing workplace."

"The world has changed dramatically in the 32 years of my professional career to date," says Bowles. "Capable engineers work hard to keep their subject matter expertise current, and they look for opportunities to broaden their expertise, as it is inevitable that they will change jobs and even industries along the way," he adds. "Change offers an opportunity to excel at something new."

Having worked with many young chemical engineers (and having hired and mentored more than a few) through the years, Bowles also notes that the specific college attended is not necessarily a major predictor of capability, nor is the GPA (within a fairly broad band). "I've worked with hundreds of engineers throughout my career, who are from schools ranging from state schools to 'name' schools and it really matters very little," he says. The correlation between the school and GPA and career opportunities "is much more tenuous than most students believe."

More importantly, he urges students to keep their grades up no matter where they go to school, because "a low enough GPA tends to be an indication of work ethic, discipline, and judgment issues, which may foreclose certain opportunities as they are making the transition from school to the workforce."

YOUR CHARACTER SPEAKS VOLUMES

At the end of the day, integrity is the cornerstone of ultimate success, says Bowles. "When you're young and you're forming your character, there's often a temptation to look for ways to cheat or cut corners, or to do just the bare minimum to get by, whether it's in course work or meeting a difficult work challenge."

"But when you commit to always behaving in a way that demonstrates high integrity, people realize it—you gain their trust, they listen to your opinion, and know they can count on you to deliver the best information, work product, or advice with no ulterior motive," he says.

"Your reputation is a really big part of what you bring to the table and it's built by delivering performance above expectations time after time after time," says Bowles, adding: "Good reputations are hard to earn but easy to lose." ■

CHEMICAL PLANT OPERATIONS

In his own words
"Stick with your chemical engineering degree
and open up a world of experiences"

ROY E. SANDERS, PE, EE, PPG INDUSTRIES (RETIRED)
AND CURRENT PROCESS SAFETY CONSULTANT

Current position:
Part-time Process Safety Consultant and Retiree from PPG Industries (Lake Charles, LA)

Education:
- BS, Chemical Engineering, Louisiana State University (Baton Rouge)
- Is a Licensed Professional Engineer (PE) and Professional Environmental Engineer (EE)

A few career highlights:
- Was instrumental in helping to define and shape the safety-oriented culture and process-safety practices of a world-class chemical facility that manufactures hundreds of tons per day of chemicals with hazardous properties (flammability, reactivity, corrosivity, toxicity, and more)
- Served as both PPG's Superintendent of Process Safety and Chairman of PPG's Technical Safety Review Board
- Currently serves on the Editorial Board of two technical journals: *Process Safety Progress* (AIChE) and *Chemical Processing* magazine
- Founding member and contributor to the AIChE's *Beacon*, a publication of the AIChE's Center for Chemical Process Safety
- Has served on the AIChE's Loss Prevention Symposium Committee since 1979, and is now an Emeritus Member
- Has published about two-dozen technical papers and, in 2015, completed the 4th Edition of his book *Chemical Process Safety: Learning from Case Histories*
- Received the Harry H. West Memorial Service Award from Texas A&M University's Mary Kay O'Connor Process Center in 2011
- Elected Fellow of the AIChE (an honor reserved for AIChE members who have proven to be strong contributing members to the institute, society, and the chemical engineering profession)

Roy Sanders spent the entirety of his 43-year professional career working at PPG Industries' chemical complex in Lake Charles, Louisiana. PPG's 1,300-acre Lake Charles facility is very large, with US$1.5 billion in assets, and between 1,500 and 2,200 employees (depending on the year).

Throughout his tenure at the company, Sanders has held various operations and technical positions, mostly related to processes that produce chlor-alkali products and chlorinated hydrocarbons. As such, he developed broad and deep knowledge and expertise related to ensuring the safe manufacture and handling of chlorine, hydrogen, sodium hydroxide, sodium chlorate, muriatic acid, sulfuric acid, vinyl chloride, ethyl chloride, trichloroethylene, perchloroethylene, and other hazardous chemicals.

"In my early days at PPG, I was primarily in operations and project design, as the company believed in rotating their people into and out of specific positions every year or so, to broaden their scope and experience," says Sanders. "I think that was very valuable. I had experience operating a caustic soda plant, a sodium chlorate plant, and an ethyl chloride plant."

"Many chemical engineers go from job to job and location to location, and they think you need to keep moving around to be successful. But sometimes you're lucky and get to stay put as you enrich your satisfaction. I stayed at my same plant location for my entire 43 years and it was always fresh and challenging as the scope changed as I went on and I gained additional responsibilities and staff over time," he says. "Clearly, I'm a living example of the fact that you can live a very comfortable life if you stick with a chemical engineering major and build your expertise from there."

A FOCUS ON PROCESS SAFETY RIGHT FROM THE START

Sanders joined the company right after graduating with his chemical engineering degree in 1965 and, at some point in the early 1970s, he was tapped to focus on loss prevention (the earlier name for process safety). "At first, I was not sure I'd be interested in that part of the job, as it was not what I'd been doing or thinking about," he says. "But it turned out to be a very good fit, and I had the good fortune to be very supported by my organization as I developed my expertise in process safety. I'm glad they asked me to because it changed the whole trajectory of my career," he adds.

He served as PPG's Superintendent of Loss Prevention for 16 years (from 1982 to 1998), and in 1999 became PPG's Compliance Team Leader (retiring with this title in 2008). With that new role, Sanders' purview expanded beyond just process safety, to also include environmental issues and personnel safety.

Throughout his career, he has conducted countless formalized Process Hazards Reviews and has investigated numerous process accidents.

Interestingly, Sanders notes that although he did not travel much while he worked in Operations at PPG, once he had developed his expertise in process safety and loss prevention, he had an opportunity (while he was still employed by PPG) to undertake some travel with the AIChE, to present detailed safety courses to chemical engineers and plant personnel. "I feel lucky to have presented interactive process safety awareness training in Bahrain, Canada, India, Saudi Arabia, Taiwan, and the Netherlands," says Sanders "This opened up a whole new world that I may never have gotten to have, had my career remained rooted in Operations at PPG," he says.

Sanders continues to present his interactive Process Safety Awareness course, entitled "What Went Wrong?" a few times a year, and he has presented numerous papers at meetings and technical symposia presented by the American Institute of Chemical Engineers (AIChE), at both AIChE meetings and AIChE's Loss Prevention Symposia, McNeese State University's Process Safety Seminars, and the Process Safety Symposia presented by Texas A&M's Mary Kay O'Connor Process Safety Center.

Since retiring, Sanders has worked as a part-time process safety consultant and lecturer, and he currently works part-time at McNeese State University (Lake Charles, Louisiana), where he coordinates four process safety seminars each year, and assists graduate students at Texas A&M's Mary Kay O'Connor Process Safety Center.

Throughout his career, Sanders has been the principal author of about two-dozen articles on chemical process safety "that offer a 'within-the-fence' look at critical practices related to operations, maintenance, and engineering," he says. In 2015, he completed the 4th Edition of his book *Chemical Process Safety: Learning from Case Histories* (Sanders 2015).

When it comes to process safety, Sanders offers this advice to young chemical engineers just starting their careers: "Don't ever become complacent and believe that process safety incidents (like accidental fires, explosions, toxic releases, or spills) can't or won't happen in your organization—because they can and they may. There is a certain vanity that many junior engineers develop, which unfortunately leads them to think such incidents cannot happen at their location or cannot happen at this time. Everything you operate, adjust, or improve can have an impact on safety, and often on the environment, too."

ON CHOOSING HIS CHEMICAL ENGINEERING MAJOR

Sanders remembers, as a high school student in New Orleans, LA, questioning whether or not he should even go to college. "I was very unsure. No one had been to college in my family and I was an immature 'class-clown wannabe' who did not fully apply myself in high school," he recalls. "I'm grateful for one particular teacher—an English teacher who also happened to be a trained pharmacist—who thought I could compete on a college campus."

"A few encouraging words to our youth is so vital," says Sanders. "It humbles me to see how just a few key words of encouragement at one time or another totally changed my life."

Sanders remembers enjoying physics and math in high school, and says: "Occasionally I also tried making wine or beer, experimented with a chemistry set, played with liquid mercury, and made gun powder. I also heard that engineers made a comfortable living, so I decided to give it a try and I put my heart and soul into my college effort."

"When I first started out in college, it was not an easy task for me, as high school was a joke to me and I did not apply myself." says Sanders. "But I pushed myself hard at a local university and channeled all my efforts in succeeding. I stuck with it and promised my family to try it one year at time. The funny thing is, when you can stick with anything for one year at a time, eventually you are able to complete your mission. After two years, I moved 90 miles and completed my chemical engineering degree at Louisiana State University, Baton Rouge—and the rest, as they say, is history."

ON THE VALUE OF PUBLIC SPEAKING

"At some point I lost my concerns about speaking in public and began presenting papers at events sponsored by the AIChE and other professional societies," says Sanders. "I used case histories with lots of images to showcase critical process safety issues. During mid-career, in the early 1980s, a rising yet little-known British chemical process safety pioneer heard one of my presentations and asked me to help him teach a two-day course in process safety for the AIChE."

Sanders continues: "At that time, Dr. Trevor A. Kletz was not yet widely recognized, but we taught a few times a year for a couple of years. Trevor had confidence in me and he mentored me and challenged me to do additional writing and teaching. I am grateful that we developed a lifelong friendship." The late Trevor A. Kletz (who passed away in 2013) is now considered to be one of the earliest, most published, and most respected "godfathers" of chemical process safety, Sanders notes.

Sanders happily discusses one achievement that has brought him the greatest personal satisfaction—when he introduced so-called "Shadow Days" to local high school students in Louisiana. Sanders enlisted plant management and employees at PPG to develop a program to periodically invite busloads of local high school students to spend a day within PPG's chemical plant. The goal was to expose students to concepts of chemical engineering during their formative years, by allowing them to observe what goes on in a chemical process plant from an employee in any number of occupations that the student could choose from. These included accountant, maintenance supervisor, chemical process operator, engineer, lab worker, safety professional, accountant, secretary, welder, painter, and so on.

"I hope each of you can find an experienced mentor who can encourage you to set your sights higher and higher—to exceed your best expectations in your career and life," says Sanders. "Only you set your own limits, so think big!" ■

CHEMICAL PLANT OPERATIONS

In his own words
Harnessing "vision and adaptability" in the pursuit of clean,
sustainable processes for fuels and chemicals

JOSEPH B. POWELL, PhD, SHELL

Current position:
Chief Scientist, Chemical Engineering for Shell (Houston, TX)

Education:
- BS, Chemical Engineering, University of Virginia
- PhD, Chemical Engineering, University of Wisconsin-Madison

A few career highlights:
- A.D. Little Award for Chemical Engineering Innovation (1998)
- Appointed Shell Chief Scientist, Chemical Engineering (2006)
- Author/editor of *Sustainable Development in the Process Industries: Cases and Impact* (2010).
- Career awards from the University of Wisconsin (2009) and the American Institute of Chemical Engineers (AIChE) Process Development Division (for service, 2012, and for practice, 2017)
- Elected to the Board of Directors of the AIChE (2016)

Since 2006, Joseph B. Powell has held the title of Chief Scientist, Chemical Engineering for the global Shell Group. "A lot of my professional life has been spent worrying about energy, including the use of fossil fuels and (later) biofuels for transportation, and producing chemical intermediates that are needed to make advantaged consumer products," he says.

In recent years, Powell has been focusing his efforts on advising long-range research efforts at Shell on the use of various engineering and business strategies to address climate change and other forms of pollution. He has also been working to increase the amount of viable wind and solar energy that's part of the energy mix produced by the company—a sign of evolution for any traditionally fossil fuels-based energy company. "We are focusing on the lofty goal of trying to create a platform of energy resources that have 'net zero' carbon emissions by 2100," he says.

While a lot of work is underway to convert certain forms of biomass (including wood chips or crop residues) into ethanol and other intermediates for the production of chemicals and plastics, the use of such renewable starting materials to produce transportation fuels (such as gasoline and diesel fuel) remains a significant challenge from a chemical engineering perspective. "Renewable feedstocks can be low cost, especially if obtained from waste streams, but collection and shipment of low-density solid biomass tends to limit the allowed scale of biorefineries that rely on them (relative to large refineries for liquid petroleum), which creates problems for the overall economics of the process," he says.

"There are also tremendous technical challenges associated with biomass conversion, including the design of catalysts that must perform in the presence of water and help to eliminate oxygen from final fuels, as well as scale-up of the process steps that are required, involving feeding and processing of solids into liquid fuels and chemicals," he adds.

For instance, "when it comes to bioprocessing of non-traditional feedstocks, selective thermo-catalytic liquefaction of biomass has proven to be more effective than Fischer-Tropsch gas-phase chemistry (breaking apart biomass to single-carbon units). As such, this approach allows for more efficient reconstruction into the molecules of interest for use in gasoline or diesel fuels," explains Powell. "Gas-phase heating or pyrolysis routes (including catalytic hydropyrolysis) are also being developed, as are new ways to pretreat biomass to allow the use of fermentation to yield desired fuel components, such as ethanol or biobased chemicals."

One of the largest research expenditures in recent years, says Powell, has been in the development of so-called second-generation biomass conversion processes, which attempt to convert less-desirable ligno-cellulosic biomass resources such as agricultural crop or forestry residues (straw, bagasse, papermill and sawmill waste, or even wood chips)—things you cannot eat, as opposed to more readily processed, edible feedstocks such as sugars or corn. "Such processes take advantage of a less-desirable but more widely available starting material, and can result in a much larger reduction of carbon footprint—in some cases, upwards of a 60-percent reduction—compared to petroleum-derived transportation fuels."

ASCENDING THE LADDER ON THE "TECHNICAL TRACK"

The role of Chief Scientist (in different technical disciplines) at Shell was created to recognize technical experts who are at the top of the proverbial technical ladder—a track that allows for growing career opportunities in parallel with the managerial track within the company. "In this role, I serve as both an internal leader, mentor, and role model, and also as an external scientific ambassador while interacting with experts outside of Shell through the pursuit of the best new technologies and more visionary, future-looking options."

"It's really important to have a long-range plan because new technology development can take years, even decades. You have to have vision and adaptability, recognizing that the future can change quite radically and that you will be forced to maneuver in new, previously unimagined spaces," he says. "I am responsible for helping the company to deliver solutions, and I have to always be aware of the changing opportunities and evolving challenges."

ADAPTING TO MEET EVOLVING CHALLENGES

Powell got into energy very early on. "When I got out of high school in 1978, and then graduate school in 1984, the United States was in the midst of an energy crisis, and there was real concern about whether we would have enough fossil fuel resources to meet our domestic energy needs for heat, power, transportation fuels, food production, and for production of chemical products," says Powell.

"Early on, I worked on synthesis gas chemistry (from the gasification of coal or reforming of natural gas, which is cheaper and more plentiful than petroleum) to produce downstream chemicals and plastics." As an undergraduate, Powell had a summer internship at FMC corporation in Princeton, NJ working on updating material balances for a coal-gasification design. "The summer internships were really fascinating experiences, from which I learned I really wanted to be that person who could work out the details of new technical challenges, connect things, and figuring out how to drive the whole concept forward," says Powell, adding: "you really need to know the fundamentals to do or create something new."

This experience motivated Powell to go on to graduate school. His graduate studies at the University of Wisconsin entailed hydroformylation and Fischer-Tropsch processes for syngas conversion, as well as advanced reactor design.

Once out of graduate school, Powell was hired by another large energy company (Exxon) to lead enhanced oil-recovery (EOR) development, using soaps or "surfactants" to flush out residual oil left behind in the reservoir after initial petroleum recovery. "It was a really exciting time to be working on lab (and eventually pilot) testing of those new technologies," says Powell. "I was trained as a reservoir engineer after my chemical engineering PhD, and I worked on this for four years."

"Eventually, with the discovery of massive oil reserves in Alaska's Prudhoe Bay—one of the largest oilfields ever discovered in North America, which was first discovered in 1968, with oil production beginning in 1977—oil prices fell and EOR was no longer a high priority," he adds.

After four years, Powell moved on to Shell to work in process development, commercializing a new and more sustainable process to make intermediates for polycarbonate resins that are used in compact disks and other consumer products. This work eventually led to the innovation and development of a new process to produce 1,3-Propanediol (PDO) as an enabler for a new polyester to replace nylon in many applications.

The latter process entailed pioneering a proprietary process based on hydroformylation of ethylene oxide. Powell was directly involved in all steps of the new process—developing the chemistry and engineering, filing patents for protection of intellectual property, designing, developing, and supervising operation of the pilot-plant, and moving the development forward to commercial deployment.

"It was such an incredibly awesome experience to start up that process commercially for Shell," says Powell. "It's an experience that doesn't compare to anything else, and in the end we had about 150 people on our team." He notes: "it was compelling and rewarding to see the impact that commercializing a new technology had on all of the people on my team, and on the well-being of their families, and to see the product out there in the marketplace." He adds: "I still have the carpet that was ultimately produced from this process in my home in Houston."

"This project, like so many others I've worked on, really underscored the importance of great team work—building great teams and fostering collaboration is absolutely essential and so rewarding, and always leads to stronger ideas, better inventions, and more workable solutions," he adds.

HOW CHEMICAL ENGINEERING DREW HIM IN

"I had a really good chemistry teacher in high school, and based on his coaching I took an achievement test in chemistry and did well, but I certainly had no idea what chemical engineers did when I was younger," says Powell. "People said it was the most difficult undergraduate major at the University of Virginia (UVA) and that kind of challenge attracted me."

"I'm really gung-ho on the field of chemical engineering as the basis for a career—you can't ever really go wrong with this major," says Powell. "And if you master the fundamentals of heat and mass transfer, material balances, chemical reactions and separations, plant design and economics, systems engineering, modeling and optimization, and more, you can readily move between established and emerging areas with ease," he adds.

Two of Powell's sons have also pursued degrees in engineering—one in mechanical engineering, and the other in chemical engineering, while a third son has a degree in information systems and computer science. "These days, many high schools are so concerned with safety that they are very cautious about doing the hands-on experiments we used to do when I was still in school," he says. "The time I spent doing chemistry or physics experiments or building computers at home with my sons—and coaching them in safety while making it all really interactive—helped them to really see what physics, chemistry, and engineering are all about, by use of tangible examples."

"My youngest son, to some extent, helped to mentor his high school chemistry teacher, because he knew the subject that well," says Powell, "while my middle son now has a PhD in robotics, thanks in part to the science kits we worked on together while he was in high school."

THE POWER OF NETWORKING

Following encouragement from an early mentor, Powell has been very active in the AIChE, and has taken on progressively larger, more-visible roles within the organization. He now serves on the AIChE's Board of Directors, and has, in the past, served as Program Chair and Co-Chair for Spring National Meetings, and for many years as programming coordinator for the Process Development Division. "At the end of the day, I would never have had the distinct honor of being named Shell's Chief Scientist, Chemical Engineering, if I had not been doing all of the external volunteer work with the AIChE (including networking, benchmarking, and broad professional exposure to chemical engineers throughout many industry sectors)," he says. "All of that made me so much more valuable to the company I was working for."

"In terms of meeting so many other chemical engineers in other industry segments and positions, and making an effort to keep expanding one's breadth and depth of experience and understanding—it never really dawned on me until I actually did it how important it was," says Powell. "And it didn't come naturally for me," he says. "Even if you're a little on the shy or introverted side, the more you do it, the more confidence you'll gain, and at the end of the day, you will be rewarded very nicely for your efforts."

"At Shell we have been working on developing the right science and engineering to pursue real breakthroughs that address the key challenges of our time, and these efforts have a great entrepreneurial feel to them—as we look within our own company, and throughout industry and academia for the best ideas and opportunities," says Powell. "It's an exciting and crucially important time to be working with connections throughout both industry and academia right now."

"I have been very fortunate to have been able to work in process development for my entire career at Shell," says Powell. "I like to joke that the 'Crisis of the Day' always finds me, but that has allowed me to work on very different and interesting things. It's gratifying to feel you're doing something that is making a real difference on issues in energy and chemistry that are impacting everyone—issues that are making front-page news." ■

REFERENCES

CEI. 2018. International Conference on Accelerated Carbonation for Environmental and Material Engineering (ACEME), www.aiche.org/cei/conferences/international-conference-on-accelerated-carbonation-environmental-and-material-engineering-aceme/2018?utm_source=Informz%20Email&utm_medium=Informz%20Email&utm_campaign=Informz&_zs=CLPqW&_zl=5QQB1 (accessed 18 January 2018).

Levin, Heather. 2017. Electronic Waste (E-Waste) Recycling and Disposal – Facts, Statistics & Solutions, www.moneycrashers.com/electronic-e-waste-recycling-disposal-facts/ (accessed 17 November 2017).

Powell, Joseph B. 2010. *Sustainable Development in the Process Industries: Cases and Impact.* New York: John Wiley & Sons, Inc.

Sanders, Roy E. 2015. *Chemical Process Safety - Learning from Case Histories,* 4th Edition. Waltham, MA: Butterworth-Heineman, an imprint of Elsevier.

Wikipedia. 2017. Electronic Waste. https://en.wikipedia.org/wiki/Electronic_waste. (accessed 17 November 2017).

Index